What Works in Conservation

2017

Edited by
William J. Sutherland, Lynn V. Dicks,
Nancy Ockendon and Rebecca K. Smith

Funded by Arcadia, Synchronicity Earth, ESRC, NERC, Natural England and Waitrose Ltd.

Contents

3. BIRD CONSERVATION

3.1 Habitat protection

Legally protect habitats for birds

Provide or retain un-harvested buffer strips

Ensure connectivity between habitat patches

3.2 Education and awareness raising

Raise awareness amongst the general public through campaigns and public information

Provide bird feeding materials to families with young children

Enhance bird taxonomy skills through higher education and training

Provide training to conservationists and land managers on bird ecology and conservation

3.3 Threat: Residential and commercial development

Angle windows to reduce bird collisions

Mark windows to reduce bird collisions

3.4 Threat: Agriculture

3.4.1 All farming systems

Plant wild bird seed or cover mixture

Provide (or retain) set-aside areas in farmland

Create uncultivated margins around intensive arable or pasture fields

Increase the proportion of natural/semi-natural habitat in the farmed landscape

Manage ditches to benefit wildlife

Pay farmers to cover the costs of conservation measures

Plant grass buffer strips/margins around arable or pasture fields

Plant nectar flower mixture/wildflower strips

Leave refuges in fields during harvest

Reduce conflict by deterring birds from taking crops (using bird scarers)

Relocate nests at harvest time to reduce nestling mortality

Use mowing techniques to reduce mortality

Control scrub on farmland

Offer per clutch payment for farmland birds

Manage hedges to benefit wildlife

Plant new hedges

Reduce conflict by deterring birds from taking crops (using repellents)

Take field corners out of management

Mark bird nests during harvest or mowing

Cross compliance standards for all subsidy payments

Food labelling schemes relating to biodiversity-friendly farming

Manage stone-faced hedge banks to benefit birds

Plant in-field trees

Protect in-field trees

Reduce field size (or maintain small fields)

Support or maintain low-intensity agricultural systems

Employ grazing in artificial grasslands/pastures

Employ grazing in natural grasslands

Employ grazing in non-grassland habitats

Manage water level in wetlands

Manually control or remove midstorey and ground-level vegetation (including mowing, chaining, cutting etc.) (forests)

Manually control or remove midstorey and ground-level vegetation (including mowing, chaining, cutting etc.) (mowing or cutting natural grasslands)

Manually control or remove midstorey and ground-level vegetation (including mowing, chaining, cutting etc.) (mowing or cutting semi-natural grasslands/ pastures)

Manually control or remove midstorey and ground-level vegetation (including mowing, chaining, cutting etc.) (shrublands)

Raise water levels in ditches or grassland

Thin trees within forests

Use prescribed burning (grasslands)

Use prescribed burning (pine forests)

Use prescribed burning (savannahs)

Use prescribed burning (shrublands)

Use selective harvesting/logging instead of clearcutting

Clearcut and re-seed forests

Coppice trees

Fertilise grasslands

Manage woodland edges for birds

Manually control or remove midstorey and ground-level vegetation (including mowing, chaining, cutting etc.) (reedbeds)

Manually control or remove midstorey and ground-level vegetation (including mowing, chaining, cutting etc.) (savannahs)

Plant trees to act as windbreaks

Plough habitats

Provide deadwood/snags in forests (adding woody debris to forests)

Remove coarse woody debris from forests

Replace non-native species of tree/shrub

Re-seed grasslands

Use environmentally sensitive flood management

Use fire suppression/control

Use greentree reservoir management

Use prescribed burning (Australian sclerophyll forest)

Use shelterwood cutting instead of clearcutting

Use variable retention management during forestry operations

Apply herbicide to mid- and understorey vegetation

Treat wetlands with herbicides

Use prescribed burning (coastal habitats)

Use prescribed burning (deciduous forests)

Protect nest trees before burning

Can nest protection increase predation of adults and chicks?

3.11.4 Reduce mortality by reducing hunting ability or changing predator behaviour

Reduce predation by translocating nest boxes

Use collar-mounted devices to reduce predation

Use supplementary feeding to reduce predation

Use aversive conditioning to reduce nest predation

3.11.5 Reduce competition with other species for food and nest sites

Reduce inter-specific competition for food by removing or controlling competitor species

Protect nest sites from competitors

Reduce competition between species by providing nest boxes

Reduce inter-specific competition for nest sites by modifying habitats to exclude competitor species

Reduce inter-specific competition for nest sites by removing competitor species (ground nesting seabirds)

Reduce inter-specific competition for nest sites by removing competitor species (songbirds)

Reduce inter-specific competition for nest sites by removing competitor species (woodpeckers)

3.11.6 Reduce adverse habitat alteration by other species

Control or remove habitat-altering mammals

Reduce adverse habitat alterations by excluding problematic species (terrestrial species)

Reduce adverse habitat alterations by excluding problematic species (aquatic species)

Remove problematic vegetation

Use buffer zones to reduce the impact of invasive plant control

3.11.7 Reduce parasitism and disease

Remove/control adult brood parasites

Remove/treat endoparasites and diseases

Alter artificial nest sites to discourage brood parasitism

Exclude or control 'reservoir species' to reduce parasite burdens

Remove brood parasite eggs from target species' nests

Remove/treat ectoparasites to increase survival or reproductive success (provide beneficial nesting material)

Remove/treat ectoparasites to increase survival or reproductive success (remove ectoparasites from feathers)

Use false brood parasite eggs to discourage brood parasitism

Remove/treat ectoparasites to increase survival or reproductive success (remove ectoparasites from nests)

3.11.8 Reduce detrimental impacts of other problematic species

Use copper strips to exclude snails from nests

Manage hedgerows to benefit wildlife (includes no spray, gap-filling and laying)
Pay farmers to cover the cost of conservation measures (as in agri-environment schemes)
Provide supplementary food for birds or mammals
Connect areas of natural or semi-natural habitat
Increase the proportion of semi-natural habitat in the farmed landscape
Make direct payments per clutch for farmland birds
Manage the agricultural landscape to enhance floral resources
Mark bird nests during harvest or mowing
Plant new hedges
Provide nest boxes for bees (solitary bees or bumblebees)
Provide nest boxes for birds
Provide other resources for birds (water, sand for bathing)
Provide refuges during harvest or mowing
Apply 'cross compliance' environmental standards linked to all subsidy payments
Implement food labelling schemes relating to biodiversity-friendly farming (organic, LEAF marque)
Introduce nest boxes stocked with solitary bees
Maintain in-field elements such as field islands and rockpiles
Manage stone-faced hedge banks to benefit wildlife
Manage woodland edges to benefit wildlife
Plant in-field trees (not farm woodland)
Protect in-field trees (includes management such as pollarding and surgery)
Provide badger gates
Provide foraging perches (e.g. for shrikes)
Provide otter holts
Provide red squirrel feeders
Reduce field size (or maintain small fields)
Restore or maintain dry stone walls
Support or maintain low intensity agricultural systems

4.2 Arable farming

Create skylark plots
Leave cultivated, uncropped margins or plots (includes 'lapwing plots')
Create beetle banks
Leave overwinter stubbles
Reduce tillage
Undersow spring cereals, with clover for example
Convert or revert arable land to permanent grassland
Create rotational grass or clover leys
Increase crop diversity
Plant cereals in wide-spaced rows
Plant crops in spring rather than autumn
Plant nettle strips
Sow rare or declining arable weeds

4.6 Threat: Agri-chemicals

Leave headlands in fields unsprayed (conservation headlands)
Reduce fertilizer, pesticide or herbicide use generally
Use organic rather than mineral fertilizers
Reduce chemical inputs in grassland management
Provide buffer strips alongside water courses (rivers and streams)
Restrict certain pesticides
Buffer in-field ponds
Make selective use of spring herbicides

4.7 Threat: Transport and service corridors

Manage land under power lines to benefit wildlife

4.8 Threat: Hunting and trapping (for pest control, food or sport)

Enforce legislation to protect birds against persecution
Provide 'sacrificial' grasslands to reduce the impact of wild geese on crops
Avoid use of lead shot
Use alerts to reduce grey partridge by-catch during shoots
Use scaring devices (e.g. gas guns) and other deterrents to reduce persecution of native species

4.9 Threat: Natural system modification

Raise water levels in ditches or grassland
Create scrapes and pools
Manage heather by swiping to simulate burning
Manage heather, gorse or grass by burning
Remove flood defence banks to allow inundation
Re-wet moorland

4.10 Threat: Invasive and other problematic species

Control predatory mammals and birds (foxes, crows, stoats and weasels)
Control scrub
Control weeds without damaging other plants in conservation areas
Protect individual nests of ground-nesting birds
Control grey squirrels
Erect predator-proof fencing around important breeding sites for waders
Manage wild deer numbers
Remove coarse fish
Control bracken
Control invasive non-native plants on farmland (such as Himalayan balsam, Japanese knotweed)
Control mink
Provide medicated grit for grouse

4.11 Threat: Education and awareness

Provide specialist advice, assistance preparing conservation plans
Provide training for land managers, farmers and farm advisers

Use soil disturbance to enhance germination (excluding scarification or ploughing)

Enhance soil compaction

5.11 Actions to improve survival and growth rate of planted trees

Prepare the ground before tree planting

Use mechanical thinning before or after planting

Fence to prevent grazing after tree planting

Use herbicide after tree planting

Use prescribed fire after tree planting

Apply insecticide to protect seedlings from invertebrates

Add lime to the soil after tree planting

Add organic matter after tree planting

Cover the ground with straw after tree planting

Improve soil quality after tree planting (excluding applying fertilizer)

Manage woody debris before tree planting

Use shading for planted trees

Use tree guards or shelters to protect planted trees

Use weed mats to protect planted trees

Water seedlings

Mechanically remove understory vegetation after tree planting

Use different planting or seeding methods

Use fertilizer after tree planting

Apply fungicide to protect seedlings from fungal diseases

Infect tree seedlings with mycorrhizae

Introduce leaf litter to forest stands

Plant a mixture of tree species to enhance the survival and growth of planted trees

Reduce erosion to increase seedling survival

Transplant trees

Use pioneer plants or crops as nurse-plants

5.12 Education and awareness raising

Provide education programmes about forests

Raise awareness amongst the general public through campaigns and public information

6. SOME ASPECTS OF CONTROL OF FRESHWATER INVASIVE SPECIES

6.1 Threat: Invasive plants

6.1.1 Floating pennywort *Hydrocotyle ranunculoides*

Chemical control using herbicides

Flame treatment

Physical removal

Combination treatment using herbicides and physical removal

Biological control using co-evolved, host-specific herbivores

Using a combination of netting and electrofishing
UV radiation

6.5 Threat: Invasive reptiles

6.5.1 Red-eared terrapin *Trachemys scripta*

Direct removal of adults
Application of a biocide
Biological control using native predators
Draining invaded waterbodies
Public education
Search and removal using sniffer dogs

6.6 Threat: Invasive amphibians

6.6.1 American bullfrog *Lithobates catesbeiana*

Biological control using native predators
Direct removal of adults
Direct removal of juveniles
Application of a biocide
Biological control of co-occurring beneficial species
Collection of egg clutches
Draining ponds
Fencing
Habitat modification
Pond destruction
Public education

7. SOME ASPECTS OF ENHANCING NATURAL PEST CONTROL

7.1 Reducing agricultural pollution

Alter the timing of insecticide use
Delay herbicide use
Incorporate parasitism rates when setting thresholds for insecticide use
Use pesticides only when pests or crop damage reach threshold levels
Convert to organic farming

7.2 All farming systems

Grow non-crop plants that produce chemicals that attract natural enemies
Use chemicals to attract natural enemies
Leave part of the crop or pasture unharvested or uncut
Plant new hedges
Use alley cropping

7.3 Arable farming

Combine trap and repellent crops in a push-pull system
Use crop rotation in potato farming systems
Create beetle banks
Incorporate plant remains into the soil that produce weed-controlling chemicals

Introduction

This book has been created to help you make decisions about practical conservation management by providing an assessment, from the available scientific evidence, of what works and what does not work in conservation. It also tells you if no evidence has been found about whether or not a conservation intervention is effective. This is the 2017 edition of *What Works in Conservation*, which was first published in 2015 and will be updated annually.

Who is *What Works in Conservation* for?

This book is for people who have to make decisions about how best to support or conserve biodiversity. These include land managers, conservationists in the public or private sector, farmers, campaigners, advisors or consultants, policymakers, researchers or people taking action to protect local wildlife. *What Works in Conservation* and the associated synopses summarize scientific evidence relevant to conservation objectives and the actions that could be taken to achieve them. *What Works in Conservation* also provides an assessment of the effectiveness of interventions based on available evidence.

We do not aim to make decisions for people, but to support decision-making by providing what evidence there is (or is not) about the effects that your planned actions could have. It is important that you read the full details of the evidence, freely available online at www.conservationevidence.com, before making any decisions about implementing an intervention.

The Conservation Evidence project

The Conservation Evidence project has four parts, all of which are available from our website conservationevidence.com:

1. An ever-expanding searchable **database of over 4,700 summaries** of previously published scientific papers, reports, reviews or systematic reviews that document the effects of interventions.

 http://dx.doi.org/10.11647/OBP.0109.09

2. **Synopses** of the evidence captured in part 1) relating to particular species groups, habitats or conservation issues. Synopses bring together the evidence for all possible interventions. Synopses are also available to purchase in printed book form.
3. ***What Works in Conservation*** provides an assessment of the effectiveness of interventions based on available evidence. It contains both the key messages from the evidence for each conservation intervention from the relevant synopses, and an assessment of the effectiveness of each intervention by expert panels.
4. An online, **open access journal** *Conservation Evidence* that publishes new pieces of research on the effects of conservation management interventions. All our papers are written by, or in conjunction with, those who carried out the conservation work and include some monitoring of its effects.

Alongside this project, the Centre for Evidence-Based Conservation (http://www.cebc.bangor.ac.uk) and the Collaboration for Environmental Evidence (http://www.environmentalevidence.org) carry out and compile systematic reviews of evidence on the effectiveness of particular conservation interventions. We recommend carrying out a systematic review, which is more comprehensive than our summaries of evidence, when decisions have to be made with particularly important consequences. Systematic reviews are included in the Conservation Evidence database.

Which conservation interventions are included?

Lists of interventions for each synopsis are developed and agreed in partnership with an advisory board made up of international conservationists and academics with expertise in the subject. We aim to include all actions that have been carried out or advised for the conservation of the specific group of species or habitat or for the specific conservation issue.

The lists of interventions are organized into categories based on the International Union for the Conservation of Nature (IUCN) classifications of direct threats and conservation actions (http://www.iucnredlist.org/technical-documents/classification-schemes). Interventions are primarily grouped according to the relevant direct threats. However, some interventions can be used in response to many different threats and so these have been grouped according to conservation action.

How we review the literature

We gather evidence by searching relevant scientific journals from volume one through to the most recent volume. Thirty general conservation journals are regularly searched by Conservation Evidence. Specialist journals are also searched for each synopsis (156 have been searched so far). We also search reports, unpublished literature and evidence provided by our advisory boards. Two of the synopses used systematic mapping exercises undertaken by, or in partnership with, other institutions. Systematic mapping uses a rigorous search protocol (involving an array of specified search terms) to retrieve studies from several scientific databases. Evidence published in languages other than English is included when it is identified. Evidence from all around the world is included in synopses. One exception is farmland conservation, which only covers northern Europe (all European countries west of Russia, but not those south of France, Switzerland, Austria, Hungary and Romania). Any apparent bias towards evidence from some regions in a particular synopsis reflects the current biases in published research papers available to Conservation Evidence.

The criteria for inclusion of studies in the Conservation Evidence database are as follows:

- A conservation intervention must have been carried out.
- The effects of the intervention must have been monitored quantitatively.

These criteria exclude studies examining the effects of specific interventions without actually doing them. For example, predictive modelling studies and studies looking at species distributions in areas with long-standing management histories (correlative studies) are excluded. Such studies can suggest that an intervention could be effective, but do not provide direct evidence of a causal relationship between the intervention and the observed biodiversity pattern.

We summarise the results of each study that are relevant to each intervention. Unless specifically stated, results reflect statistical tests performed on the data within the papers.

What does *What Works in Conservation* include?

What Works in Conservation includes **only the key messages from each synopsis**, which provide a rapid overview of the evidence. These messages are condensed from the summary text for each intervention within each synopsis. **For the full text and references see www.conservationevidence.com**

Panels of experts have assessed the collated evidence for each intervention to determine effectiveness, certainty of the evidence and, in most cases, whether there are negative side-effects (harms). Using these assessments, interventions are categorized based on a combination of effectiveness (the size of benefit or harm) and certainty (the strength of the evidence). The following categories are used: Beneficial, Likely to be beneficial, Trade-off between benefit and harms, Unknown effectiveness, Unlikely to be beneficial, Likely to be ineffective or harmful (for more details see below).

Expert assessment of the evidence

The average of several experts' opinions is likely to be a more reliable and accurate assessment than the opinion of a single expert. We therefore ask a panel of experts to use their judgement to assess whether evidence within the synopsis indicates that an intervention is effective or not. They are also asked to assess how certain they are of the effectiveness given the quality of evidence available for that intervention (certainty of the evidence). Negative side-effects described in the collated evidence are also assessed (harms). They base their assessment solely on the evidence in the synopsis. We use a modified Delphi method to quantify the effectiveness and certainty of evidence of each intervention, based on the summarized evidence. The Delphi method is a structured process that involves asking a panel of experts to state their individual opinion on a subject by scoring anonymously. They can then revise their own scores after seeing a summary of scores and comments from the rest of the panel. Final scores are then collated. Scores and comments are kept anonymous throughout the process so that participants are not overly influenced by any single member of the panel.

For each intervention, experts are asked to read the summarized evidence in the synopsis and then score to indicate their assessment of the following:

Effectiveness: 0 = no effect, 100% = always effective.

Certainty of the evidence: 0 = no evidence, 100% = high quality evidence; complete certainty. This is certainty of effectiveness of intervention, not of harms.

Harms: 0 = none, 100% = major negative side-effects to the group of species/ habitat of concern.

Categorization of interventions

After one or two rounds of initial scoring, interventions are categorized by their effectiveness, as assessed by the expert panel. The median score from all the experts' assessments is calculated for the effectiveness, certainty and harms for each intervention. Categorization is based on these median values i.e. on a combination of the size of the benefit and harm and the strength of the evidence. The table and figure overleaf show how interventions are categorized using the median scores. There is an important distinction between lack of benefit and lack of evidence of benefit.

Once interventions are categorized, experts are given the chance to object if they believe an intervention has been categorized incorrectly. Interventions that receive a specified number (depending on the size of the panel) of strong objections from experts are re-scored by the expert panel and re-categorized accordingly. Experts did not see the categories for the farmland synopsis or for the 'Reduce predation by other species' section of the bird synopsis and so those categories are based on the second round of scoring.

How to use *What Works in Conservation*

Please remember that the categories provided in this book are meant as a guide and a starting point in assessing the effectiveness of conservation interventions. The assessments are based on the available evidence for the target group of species for each intervention and may therefore refer to different species or habitat to the one(s) you are considering. Before making any decisions about implementing interventions it is vital that you read the more detailed accounts of the evidence, in order to assess their relevance to your species or system. Full details of the evidence are available at www.conservationevidence.com.

There may also be significant negative side-effects on the target groups or other species or communities that have not been identified in our assessment. A lack of evidence means that we have been unable to assess whether or not an intervention is effective or has any harmful impacts.

Table of categories of effectiveness

Category	Description	General criteria	Thresholds
Beneficial	Effectiveness has been demonstrated by clear evidence. Expectation of harms is small compared with the benefits	High median benefit score High median certainty score Low median harm score	Effectiveness: >60% Certainty: >60% Harm: <20%
Likely to be beneficial	Effectiveness is less well established than for those listed under 'beneficial' **OR** There is clear evidence of medium effectiveness	High benefit score Lower certainty score Low harm score **OR** Medium benefit score High certainty score Low harm score	Effectiveness: >60% Certainty: 40–60% Harm: <20% **OR** Effectiveness: 40–60% Certainty: ≥40% Harm: <20%
Trade-off between benefit and harms	Interventions for which practitioners must weigh up the beneficial and harmful effects according to individual circumstances and priorities	Medium benefit and medium harm scores **OR** High benefit and high harm scores High certainty score	Effectiveness: ≥40% Certainty: ≥40% Harm: ≥20%
Unknown effectiveness (limited evidence)	Currently insufficient data, or data of inadequate quality	Low certainty score	Effectiveness: Any Certainty: <40% Harm: Any
Unlikely to be beneficial	Lack of effectiveness is less well established than for those listed under 'likely to be ineffective or harmful'	Low benefit score Medium certainty score and/or some variation between experts	Effectiveness: <40% Certainty: 40–60% Harm: <20%
Likely to be ineffective or harmful	Ineffectiveness or harmfulness has been demonstrated by clear evidence	Low benefit score High certainty score (regardless of harms) **OR** Low benefit score High harm score (regardless of certainty of effectiveness)	Effectiveness: <40% Certainty: >60% Harm: Any **OR** Effectiveness: <40% Certainty: ≥ 40% Harm: ≥20%

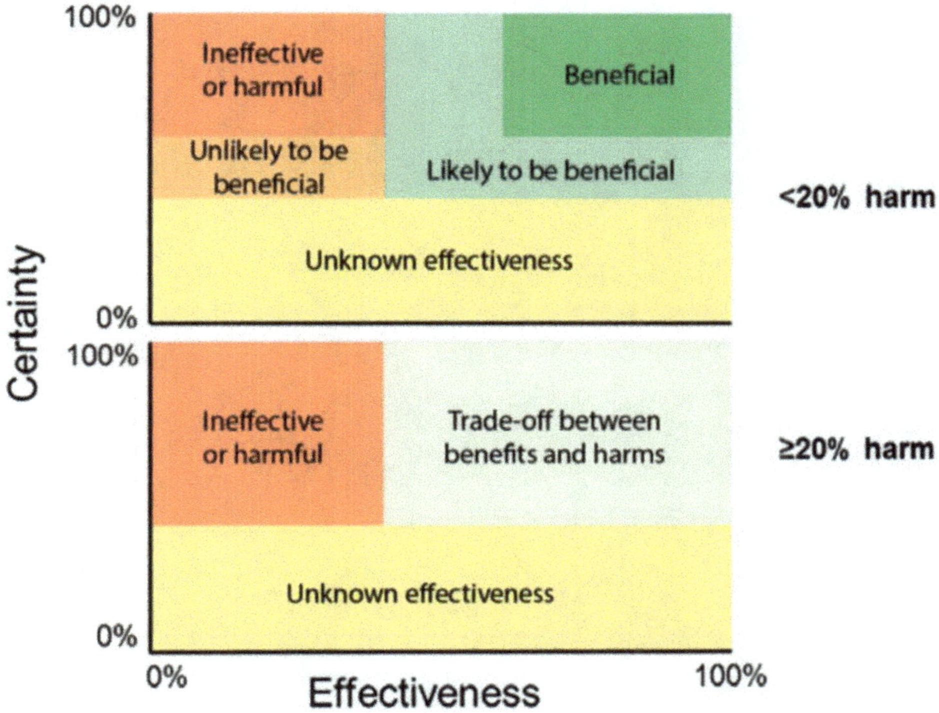

Categories of effectiveness based on a combination of effectiveness (the size of the benefit and harm) and certainty (the strength of the evidence). The top graph refers to interventions with harms <20% and the bottom graph to interventions with harms ≥20%.

1. AMPHIBIAN CONSERVATION

Rebecca K. Smith, Helen Meredith & William J. Sutherland

Expert assessors

Ariadne Angulo, Co-Chair of the Amphibian Specialist Group, Peru
Robert Brodman, Saint Joseph's College, Indiana, USA
Andrew Cunningham, Institute of Zoology, Zoological Society of London, UK
Jeff Dawson, Durrell Wildlife Conservation Trust, UK
Rob Gandola, University of Southampton, UK
Jaime García Moreno, International Union for Conservation of Nature, The Netherlands
Trent Garner, Institute of Zoology, Zoological Society of London, UK
Richard Griffiths, Durrell Institute of Conservation and Ecology, UK
Sergei Kuzmin, Russian Academy of Sciences
Michael Lanoo, Indiana University, USA
Michael Lau, WWF-Hong Kong
James Lewis, Amphibian Survival Alliance/Global Wildlife Conservation, USA
An Martel, Ghent University, Belgium
LeGrand Nono Gonwouo, Cameroon Herpetology-Conservation Biology Foundation
Deanna Olson, US Forest Service
Timo Paasikunnas, Curator of Conservation at Helsinki Zoo, Finland
Frank Pasmans, Ghent University, Belgium
Silviu Petrovan, Froglife, UK
Carlos Martínez Rivera, Philadelphia Zoo, USA
Gonçalo Rosa, Institute of Zoology, Zoological Society of London, UK
David Sewell, Durrell Institute of Conservation and Ecology, UK
Rebecca K. Smith, University of Cambridge, UK
Ben Tapley, Herpetology Department, Zoological Society of London, UK
Jeanne Tarrant, Endangered Wildlife Trust, South Africa
Karthikeyan Vasudevan, Wildlife Institute of India
Victor Wasonga, National Museums of Kenya
Ché Weldon, North-West University, South Africa
Sally Wren, Amphibian Specialist Group Programme Officer, New Zealand

Scope of assessment: for native wild amphibian species across the world.

Assessed: 2014.

Effectiveness measure is the median % score for effectiveness.

Certainty measure is the median % certainty of evidence for effectiveness, determined by the quantity and quality of the evidence in the synopsis.

Harm measure is the median % score for negative side-effects to the group of species of concern.

 http://dx.doi.org/10.11647/OBP.0109.01

This book is meant as a guide to the evidence available for different conservation interventions and as a starting point in assessing their effectiveness. The assessments are based on the available evidence for the target group of species for each intervention. The assessment may therefore refer to different species or habitat to the one(s) you are considering. Before making any decisions about implementing interventions it is vital that you read the more detailed accounts of the evidence in order to assess their relevance for your study species or system.

Full details of the evidence are available at
www.conservationevidence.com

There may also be significant negative side-effects on the target groups or other species or communities that have not been identified in this assessment.

A lack of evidence means that we have been unable to assess whether or not an intervention is effective or has any harmful impacts.

1.1 Threat: Residential and commercial development

Based on the collated evidence, what is the current assessment of the effectiveness of interventions for residential and commercial development?	
Unknown effectiveness (limited evidence)	• Legal protection of species
No evidence found (no assessment)	• Protect brownfield or ex-industrial sites • Restrict herbicide, fungicide and pesticide use on and around ponds on golf courses

Unknown effectiveness (limited evidence)

Legal protection of species

Three reviews, including one systematic review, in the Netherlands and UK found that legal protection of amphibians was not effective at protecting populations during development. Two reviews found that the number of great crested newt mitigation licences issued in England and Wales increased over 10 years. *Assessment: unknown effectiveness — limited evidence (effectiveness 10%; certainty 35%; harms 7%).*

http://www.conservationevidence.com/actions/779

No evidence found (no assessment)

We have captured no evidence for the following interventions:

- Protect brownfield or ex-industrial sites
- Restrict herbicide, fungicide and pesticide use on and around ponds on golf courses

1.2 Threat: Agriculture

1.2.1 Engage farmers and other volunteers

Based on the collated evidence, what is the current assessment of the effectiveness of interventions for engaging farmers and other volunteers?	
Likely to be beneficial	• Engage landowners and other volunteers to manage land for amphibians • Pay farmers to cover the costs of conservation measures

Likely to be beneficial

Engage landowners and other volunteers to manage land for amphibians

Three studies, including one replicated and one controlled study, in Estonia, Mexico and Taiwan found that engaging landowners and other volunteers in habitat management increased amphibian populations and axolotl weight. Six studies in Estonia, the USA and UK found that up to 41,000 volunteers were engaged in habitat restoration programmes for amphibians and restored up to 1,023 ponds or 11,500 km2 of habitat. *Assessment: likely to be beneficial (effectiveness 70%; certainty 55%; harms 5%).*

http://www.conservationevidence.com/actions/777

Pay farmers to cover the costs of conservation measures

Four of five studies, including two replicated studies, in Denmark, Sweden and Taiwan found that payments to farmers increased amphibian populations, numbers of species or breeding habitat. One found that amphibian habitat was not maintained. *Assessment: likely to be beneficial (effectiveness 70%; certainty 53%; harms 10%).*

http://www.conservationevidence.com/actions/818

1.2.2 Terrestrial habitat management

Based on the collated evidence, what is the current assessment of the effectiveness of interventions for terrestrial habitat management in agricultural systems?	
Unknown effectiveness (limited evidence)	• Manage cutting regime • Manage grazing regime
No evidence found (no assessment)	• Maintain or restore hedges • Plant new hedges • Reduced tillage

Manage silviculture practices in plantations

Studies investigating the effects of silviculture practices are discussed in 'Threat: Biological resource use — Logging and wood harvesting'.

Unknown effectiveness (limited evidence)

Manage cutting regime

One before-and-after study in Australia found that restoration that included reduced mowing increased numbers of frog species. *Assessment for 'Change mowing regime' from 'Habitat restoration and creation' section: unknown effectiveness — limited evidence (effectiveness 50%; certainty 30%; harms 0%).*

http://www.conservationevidence.com/actions/788

Manage grazing regime

Two studies, including one replicated, controlled study, in the UK and USA found that grazed plots had lower numbers of toads than ungrazed plots and that grazing, along with burning, decreased numbers of amphibian species. Five studies, including four replicated studies, in Denmark, Estonia and the UK found that habitat management that included reintroduction of grazing maintained or increased toad populations. *Assessment: unknown effectiveness — limited evidence (effectiveness 45%; certainty 39%; harms 10%).*

http://www.conservationevidence.com/actions/780

No evidence found (no assessment)

We have captured no evidence for the following interventions:

- Maintain or restore hedges
- Plant new hedges
- Reduced tillage

1.2.3 Aquatic habitat management

Based on the collated evidence, what is the current assessment of the effectiveness of interventions for aquatic habitat management in agricultural systems?	
Likely to be beneficial	• Manage ditches
Likely to be ineffective or harmful	• Exclude domestic animals or wild hogs from ponds by fencing

Likely to be beneficial

Manage ditches

One controlled, before-and-after study in the UK found that managing ditches increased toad numbers. One replicated, site comparison study in the Netherlands found that numbers of amphibians and species were higher in ditches managed under agri-environment schemes compared to those managed conventionally. *Assessment: likely to be beneficial (effectiveness 71%; certainty 60%; harms 0%).*

http://www.conservationevidence.com/actions/749

Likely to be ineffective or harmful

Exclude domestic animals or wild hogs from ponds by fencing

Four replicated studies, including one randomized, controlled, before-and-after study, in the USA found that excluding livestock from streams or ponds did not increase overall numbers of amphibians, species, eggs or larval survival, but did increase larval and metamorph abundance. One before-and-after study in the UK found that pond restoration that included livestock exclusion increased pond use by breeding toads. *Assessment: likely to be ineffective or harmful (effectiveness 31%; certainty 50%; harms 25%).*

http://www.conservationevidence.com/actions/746

1.3 Threat: Energy production and mining

Based on the collated evidence, what is the current assessment of the effectiveness of interventions for energy production and mining?	
Unknown effectiveness (limited evidence)	• Artificially mist habitat to keep it damp

Unknown effectiveness (limited evidence)

Artificially mist habitat to keep it damp

One before-and-after study in Tanzania found that installing a sprinkler system to mitigate against a reduction of river flow did not maintain a population of Kihansi spray toads. *Assessment: unknown effectiveness — limited evidence (effectiveness 24%; certainty 20%; harms 0%).*

http://www.conservationevidence.com/actions/755

1.4 Threat: Transportation and service corridors

Based on the collated evidence, what is the current assessment of the effectiveness of interventions for transportation and service corridors?	
Likely to be beneficial	• Close roads during seasonal amphibian migration • Modify gully pots and kerbs
Trade-off between benefit and harms	• Install barrier fencing along roads • Install culverts or tunnels as road crossings
Unknown effectiveness (limited evidence)	• Use signage to warn motorists
Unlikely to be beneficial	• Use humans to assist migrating amphibians across roads

Likely to be beneficial

Close roads during seasonal amphibian migration

Two studies, including one replicated study, in Germany found that road closure sites protected large numbers of amphibians from mortality during breeding migrations. *Assessment: likely to be beneficial (effectiveness 85%; certainty 50%; harms 0%).*

http://www.conservationevidence.com/actions/842

Modify gully pots and kerbs

One before-and-after study in the UK found that moving gully pots 10 cm away from the kerb decreased the number of great crested newts that fell in by 80%. *Assessment: likely to be beneficial (effectiveness 80%; certainty 40%; harms 0%).*

http://www.conservationevidence.com/actions/782

Trade-off between benefit and harms

Install barrier fencing along roads

Seven of eight studies, including one replicated and two controlled studies, in Germany, Canada and the USA found that barrier fencing with culverts decreased amphibian road deaths, in three cases depending on fence design. One study found that few amphibians were diverted by barriers. *Assessment: trade-offs between benefits and harms (effectiveness 65%; certainty 68%; harms 23%).*

http://www.conservationevidence.com/actions/756

Install culverts or tunnels as road crossings

Thirty-two studies investigated the effectiveness of installing culverts or tunnels as road crossings for amphibians. Six of seven studies, including three replicated studies, in Canada, Europe and the USA found that installing culverts or tunnels decreased amphibian road deaths. One found no effect on road deaths. Fifteen of 24 studies, including one review, in Australia, Canada, Europe and the USA found that tunnels were used by amphibians. Four found mixed effects depending on species, site or culvert type. Five found that culverts were not used or were used by less than 10% of amphibians. Six studies, including one replicated, controlled study, in Canada, Europe and the USA investigated the use of culverts with flowing water. Two found that they were used by amphibians. Three found that they were rarely or not used. Certain culvert designs were found not to be suitable for amphibians. *Assessment: trade-offs between benefits and harms (effectiveness 60%; certainty 75%; harms 25%).*

http://www.conservationevidence.com/actions/884

Unknown effectiveness (limited evidence)

Use signage to warn motorists

One study in the UK found that despite warning signs and human assistance across roads, some toads were still killed on roads. *Assessment: unknown effectiveness — limited evidence (effectiveness 10%; certainty 10%; harms 0%).*

http://www.conservationevidence.com/actions/841

Unlikely to be beneficial

Use humans to assist migrating amphibians across roads

Three studies, including one replicated study, in Italy and the UK found that despite assisting toads across roads during breeding migrations, toads were still killed on roads and 64–70% of populations declined. Five studies in Germany, Italy and the UK found that large numbers of amphibians were moved across roads by up to 400 patrols. *Assessment: unlikely to be beneficial (effectiveness 35%; certainty 40%; harms 3%).*

http://www.conservationevidence.com/actions/784

1.5 Threat: Biological resource use

1.5.1 Hunting and collecting terrestrial animals

Based on the collated evidence, what is the current assessment of the effectiveness of interventions for hunting and collecting terrestrial animals?	
Likely to be beneficial	• Reduce impact of amphibian trade
Unknown effectiveness (limited evidence)	• Use legislative regulation to protect wild populations
No evidence found (no assessment)	• Commercially breed amphibians for the pet trade • Use amphibians sustainably

Likely to be beneficial

Reduce impact of amphibian trade

One review found that reducing trade through legislation allowed frog populations to recover from over-exploitation. *Assessment: likely to be beneficial (effectiveness 76%; certainty 40%; harms 0%).*

http://www.conservationevidence.com/actions/824

Unknown effectiveness (limited evidence)

Use legislative regulation to protect wild populations

One review found that legislation to reduce trade resulted in the recovery of frog populations. One study in South Africa found that the number of permits issued for scientific and educational use of amphibians increased from 1987 to 1990. *Assessment: unknown effectiveness — limited evidence (effectiveness 60%; certainty 30%; harms 5%).*

http://www.conservationevidence.com/actions/785

No evidence found (no assessment)

We have captured no evidence for the following interventions:

- Commercially breed amphibians for the pet trade
- Use amphibians sustainably

1.5.2 Logging and wood harvesting

Based on the collated evidence, what is the current assessment of the effectiveness of interventions for logging and wood harvest?	
Likely to be beneficial	• Retain riparian buffer strips during timber harvest • Use shelterwood harvesting instead of clearcutting
Trade-off between benefit and harms	• Leave coarse woody debris in forests
Unknown effectiveness (limited evidence)	• Use patch retention harvesting instead of clearcutting
Unlikely to be beneficial	• Leave standing deadwood/snags in forests • Use leave-tree harvesting instead of clearcutting
Likely to be ineffective or harmful	• Harvest groups of trees instead of clearcutting • Thin trees within forests

Likely to be beneficial

Retain riparian buffer strips during timber harvest

Six replicated and/or controlled studies in Canada and the USA compared amphibian numbers following clearcutting with or without riparian buffer strips. Five found mixed effects and one found that abundance was higher with riparian buffers. Two of four replicated studies, including one randomized, controlled, before-and-after study, in Canada and the USA found that numbers of species and abundance were greater in wider buffer strips. Two found no effect of buffer width. *Assessment: likely to be beneficial (effectiveness 50%; certainty 61%; harms 10%).*

http://www.conservationevidence.com/actions/747

Use shelterwood harvesting instead of clearcutting

Three studies, including two randomized, replicated, controlled, before-and-after studies, in the USA found that compared to clearcutting, shelterwood harvesting resulted in higher or similar salamander abundance. One meta-analysis of studies in North America found that partial harvest, which included shelterwood harvesting, resulted in smaller reductions in salamander populations than clearcutting. *Assessment: likely to be beneficial (effectiveness 40%; certainty 57%; harms 10%).*

http://www.conservationevidence.com/actions/851

Trade-off between benefit and harms

Leave coarse woody debris in forests

Two replicated, controlled studies in the USA found that abundance was similar in clearcuts with woody debris retained or removed for eight of nine amphibian species, but that the overall response of amphibians was more negative where woody debris was retained. Two replicated, controlled studies in the USA and Indonesia found that the removal of coarse woody debris from standing forest did not affect amphibian diversity or overall amphibian abundance, but did reduce species richness. One replicated, controlled study in the USA found that migrating amphibians used clearcuts where woody debris was retained more than where it was removed. One replicated, site comparison study in the USA found that within clearcut forest, survival of juvenile amphibians was significantly higher within piles of woody debris than in open areas. *Assessment: trade-offs between benefits and harms (effectiveness 40%; certainty 60%; harms 26%).*

http://www.conservationevidence.com/actions/843

Unknown effectiveness (limited evidence)

Use patch retention harvesting instead of clearcutting

We found no evidence for the effect of retaining patches of trees rather than clearcutting on amphibian populations. One replicated study in Canada found that although released red-legged frogs did not move towards retained tree patches, large patches were selected more and moved out of less than small patches. *Assessment: unknown effectiveness — limited evidence (effectiveness 20%; certainty 25%; harms 0%).*

http://www.conservationevidence.com/actions/847

Unlikely to be beneficial

Leave standing deadwood/snags in forests

One randomized, replicated, controlled, before-and-after study in the USA found that compared to total clearcutting, leaving dead and wildlife trees did not result in higher abundances of salamanders. One randomized, replicated, controlled study in the USA found that numbers of amphibians and species were similar with removal or creation of dead trees within forest. *Assessment: unlikely to be beneficial (effectiveness 5%; certainty 58%; harms 2%).*

http://www.conservationevidence.com/actions/845

Use leave-tree harvesting instead of clearcutting

Two studies, including one randomized, replicated, controlled, before-and-after study, in the USA found that compared to clearcutting, leaving a low density of trees during harvest did not result in higher salamander abundance. *Assessment: unlikely to be beneficial (effectiveness 10%; certainty 48%; harms 11%).*

http://www.conservationevidence.com/actions/846

Likely to be ineffective or harmful

Harvest groups of trees instead of clearcutting

Three studies, including two randomized, replicated, controlled, before-and-after studies, in the USA found that harvesting trees in small groups resulted in similar amphibian abundance to clearcutting. One meta-analysis and one randomized, replicated, controlled, before-and-after study in North America and the USA found that harvesting, which included harvesting groups of trees, resulted in smaller reductions in salamander populations than clearcutting. *Assessment: likely to be ineffective or harmful (effectiveness 33%; certainty 60%; harms 23%).*

http://www.conservationevidence.com/actions/844

Thin trees within forests

Six studies, including five replicated and/or controlled studies, in the USA compared amphibians in thinned to unharvested forest. Three found that thinning had mixed effects and one found no effect on abundance. One found that amphibian abundance increased following thinning but the body condition of ensatina salamanders decreased. One found a negative overall response of amphibians. Four studies, including two replicated, controlled studies, in the USA compared amphibians in thinned to clearcut forest. Two found that thinning had mixed effects on abundance and two found higher amphibian abundance or a less negative overall response of amphibians following thinning. One meta-analysis of studies in North America found that partial harvest, which included thinning, decreased salamander populations, but resulted in smaller reductions than clearcutting. *Assessment: likely to be ineffective or harmful (effectiveness 35%; certainty 60%; harms 40%).*

http://www.conservationevidence.com/actions/852

1.6 Threat: Human intrusions and disturbance

Based on the collated evidence, what is the current assessment of the effectiveness of interventions for human intrusions and disturbance?	
No evidence found (no assessment)	• Use signs and access restrictions to reduce disturbance

No evidence found (no assessment)

We have captured no evidence for the following intervention:

- Use signs and access restrictions to reduce disturbance

1.7 Threat: Natural system modifications

Based on the collated evidence, what is the current assessment of the effectiveness of interventions for natural system modifications?	
Beneficial	• Regulate water levels
Unknown effectiveness (limited evidence)	• Mechanically remove mid-storey or ground vegetation
Likely to be ineffective or harmful	• Use herbicides to control mid-storey or ground vegetation • Use prescribed fire or modifications to burning regime: forests • Use prescribed fire or modifications to burning regime: grassland

Beneficial

Regulate water levels

Three studies, including one replicated, site comparison study, in the UK and USA found that maintaining pond water levels, in two cases with other habitat management, increased or maintained amphibian populations or increased breeding success. One replicated, controlled study in Brazil found that keeping rice fields flooded after harvest did not change amphibian abundance or numbers of species, but changed species composition. One replicated, controlled study in the USA found that draining ponds increased abundance and numbers of amphibian species. *Assessment: beneficial (effectiveness 70%; certainty 65%; harms 10%).*

http://www.conservationevidence.com/actions/833

Unknown effectiveness (limited evidence)

Mechanically remove mid-storey or ground vegetation

One randomized, replicated, controlled study in the USA found that mechanical understory reduction increased numbers of amphibian species, but not amphibian abundance. *Assessment: unknown effectiveness — limited evidence (effectiveness 40%; certainty 30%; harms 0%).*

http://www.conservationevidence.com/actions/781

Likely to be ineffective or harmful

Use herbicides to control mid-storey or ground vegetation

Three studies, including two randomized, replicated, controlled studies, in the USA found that understory removal using herbicide had no effect or negative effects on amphibian abundance. One replicated, site comparison study in Canada found that following logging, abundance was similar or lower in stands with herbicide treatment and planting compared to those left to regenerate naturally. *Assessment: likely to be ineffective or harmful (effectiveness 10%; certainty 50%; harms 50%).*

http://www.conservationevidence.com/actions/778

Use prescribed fire or modifications to burning regime (forests)

Eight of 15 studies, including three randomized, replicated, controlled studies, in Australia, North America and the USA found no effect of prescribed forest fires on amphibian abundance or numbers of species. Four found that fires had mixed effects on abundance. Four found that abundance, numbers of species or hatching success increased and one that abundance decreased. *Assessment: likely to be ineffective or harmful (effectiveness 30%; certainty 58%; harms 40%).*

http://www.conservationevidence.com/actions/877

Use prescribed fire or modifications to burning regime (grassland)

Two of three studies, including one replicated, before-and-after study, in the USA and Argentina found that prescribed fires in grassland decreased amphibian abundance or numbers of species. One found that spring, but not autumn or winter burns in grassland, decreased abundance. *Assessment: likely to be ineffective or harmful (effectiveness 10%; certainty 40%; harms 70%).*

http://www.conservationevidence.com/actions/862

1.8 Threat: Invasive and other problematic species

1.8.1 Reduce predation by other species

Based on the collated evidence, what is the current assessment of the effectiveness of interventions for reducing predation by other species?	
Beneficial	• Remove or control fish by drying out ponds
Likely to be beneficial	• Remove or control fish population by catching • Remove or control invasive bullfrogs • Remove or control invasive viperine snake • Remove or control mammals
Trade-off between benefit and harms	• Remove or control fish using Rotenone
Unknown effectiveness (limited evidence)	• Exclude fish with barriers
No evidence found (no assessment)	• Encourage aquatic plant growth as refuge against fish predation • Remove or control non-native crayfish

Beneficial

● Remove or control fish by drying out ponds

One before-and-after study in the USA found that draining ponds to eliminate fish increased numbers of amphibian species. Four studies, including one review, in Estonia, the UK and USA found that pond drying to eliminate fish, along with other management activities, increased amphibian abundance, numbers of species and breeding success. *Assessment: beneficial (effectiveness 80%; certainty 66%; harms 3%).*

http://www.conservationevidence.com/actions/826

Likely to be beneficial

Remove or control fish population by catching

Four of six studies, including two replicated, controlled studies, in Sweden, the USA and UK found that removing fish by catching them increased amphibian abundance, survival and recruitment. Two found no significant effect on newt populations or toad breeding success. *Assessment: likely to be beneficial (effectiveness 50%; certainty 52%; harms 0%).*

http://www.conservationevidence.com/actions/827

Remove or control invasive bullfrogs

Two studies, including one replicated, before-and-after study, in the USA and Mexico found that removing American bullfrogs increased the size and range of frog populations. One replicated, before-and-after study in the USA found that following bullfrog removal, frogs were found out in the open more. *Assessment: likely to be beneficial (effectiveness 79%; certainty 60%; harms 0%).*

http://www.conservationevidence.com/actions/825

Remove or control invasive viperine snake

One before-and-after study in Mallorca found that numbers of Mallorcan midwife toad larvae increased after intensive, but not less intensive, removal of viperine snakes. *Assessment: likely to be beneficial (effectiveness 50%; certainty 40%; harms 0%).*

http://www.conservationevidence.com/actions/830

Remove or control mammals

One controlled study in New Zealand found that controlling rats had no significant effect on numbers of Hochstetter's frog. Two studies, one of which was controlled, in New Zealand found that predator-proof enclosures enabled or increased survival of frog species. *Assessment: likely to be beneficial (effectiveness 50%; certainty 40%; harms 0%).*

http://www.conservationevidence.com/actions/839

Trade-off between benefit and harms

Remove or control fish using Rotenone

Three studies, including one replicated study, in Sweden, the UK and USA found that eliminating fish using rotenone increased numbers of amphibians, amphibian species and recruitment. One review in Australia, the UK and USA found that fish control that included using rotenone increased breeding success. Two replicated studies in Pakistan and the UK found that rotenone use resulted in frog deaths and negative effects on newts. *Assessment: trade-offs between benefits and harms (effectiveness 65%; certainty 60%; harms 52%).*

http://www.conservationevidence.com/actions/828

Unknown effectiveness (limited evidence)

Exclude fish with barriers

One controlled study in Mexico found that excluding fish using a barrier increased weight gain of axolotls. *Assessment: unknown effectiveness — limited evidence (effectiveness 30%; certainty 20%; harms 0%).*

http://www.conservationevidence.com/actions/829

No evidence found (no assessment)

We have captured no evidence for the following interventions:

- Encourage aquatic plant growth as refuge against fish predation
- Remove or control non-native crayfish

1.8.2 Reduce competition with other species

Based on the collated evidence, what is the current assessment of the effectiveness of interventions for reducing competition with other species?	
Unknown effectiveness (limited evidence)	• Reduce competition from native amphibians • Remove or control invasive Cuban tree frogs
No evidence found (no assessment)	• Remove or control invasive cane toads

Unknown effectiveness (limited evidence)

Reduce competition from native amphibians

One replicated, site comparison study in the UK found that common toad control did not increase natterjack toad populations. *Assessment: unknown effectiveness — limited evidence (effectiveness 10%; certainty 23%; harms 0%).*

http://www.conservationevidence.com/actions/821

Remove or control invasive Cuban tree frogs

One before-and-after study in the USA found that removal of invasive Cuban tree frogs increased numbers of native frogs. *Assessment: unknown effectiveness — limited evidence (effectiveness 65%; certainty 30%; harms 0%).*

http://www.conservationevidence.com/actions/822

No evidence found (no assessment)

We have captured no evidence for the following interventions:

- Remove or control invasive cane toads

1.8.3 Reduce adverse habitat alteration by other species

Based on the collated evidence, what is the current assessment of the effectiveness of interventions for reducing adverse habitat alteration by other species?	
Likely to be beneficial	• Control invasive plants
No evidence found (no assessment)	• Prevent heavy usage/exclude wildfowl from aquatic habitat

Likely to be beneficial

Control invasive plants

One before-and-after study in the UK found that habitat and species management that included controlling swamp stonecrop, increased a population of natterjack toads. One replicated, controlled study in the USA found that more Oregon spotted frogs laid eggs in areas where invasive reed canarygrass was mown. *Assessment: likely to be beneficial (effectiveness 60%; certainty 47%; harms 0%).*

http://www.conservationevidence.com/actions/823

No evidence found (no assessment)

We have captured no evidence for the following intervention:

- Prevent heavy usage/exclude wildfowl from aquatic habitat

1.8.4 Reduce parasitism and disease – chytridiomycosis

Based on the collated evidence, what is the current assessment of the effectiveness of interventions for reducing chytridiomycosis?	
Likely to be beneficial	• Use temperature treatment to reduce infection
Trade-off between benefit and harms	• Use antifungal treatment to reduce infection
Unknown effectiveness (limited evidence)	• Add salt to ponds • Immunize amphibians against infection • Remove the chytrid fungus from ponds • Sterilize equipment when moving between amphibian sites • Treating amphibians in the wild or pre-release • Use gloves to handle amphibians
Unlikely to be beneficial	• Use antibacterial treatment to reduce infection • Use antifungal skin bacteria or peptides to reduce infection
No evidence found (no assessment)	• Use zooplankton to remove zoospores

Likely to be beneficial

Use temperature treatment to reduce infection

Four of five studies, including four replicated, controlled studies, in Australia, Switzerland and the USA found that increasing enclosure or water temperature to 30–37°C for over 16 hours cured amphibians of chytridiomycosis. One found that treatment did not cure frogs. *Assessment: likely to be beneficial (effectiveness 60%; certainty 70%; harms 10%).*

http://www.conservationevidence.com/actions/770

Trade-off between benefit and harms

Use antifungal treatment to reduce infection

Twelve of 16 studies, including four randomized, replicated, controlled studies, in Europe, Australia, Tasmania, Japan and the USA found that antifungal treatment cured or increased survival of amphibians with chytridiomycosis. Four studies found that treatments did not cure chytridiomycosis, but did reduce infection levels or had mixed results. Six of the eight studies testing treatment with itraconazole found that it was effective at curing chytridiomycosis. One found that it reduced infection levels and one found mixed effects. Six studies found that specific fungicides caused death or other negative side effects in amphibians. *Assessment: trade-offs between benefits and harms (effectiveness 71%; certainty 70%; harms 50%).*

http://www.conservationevidence.com/actions/882

Unknown effectiveness (limited evidence)

Add salt to ponds

One study in Australia found that following addition of salt to a pond containing the chytrid fungus, a population of green and golden bell frogs remained free of chytridiomycosis for over six months. *Assessment: unknown effectiveness — limited evidence (effectiveness 41%; certainty 25%; harms 50%).*

http://www.conservationevidence.com/actions/762

Immunize amphibians against infection

One randomized, replicated, controlled study in the USA found that vaccinating mountain yellow-legged frogs with formalin-killed chytrid fungus did not significantly reduce chytridiomycosis infection rate or mortality. *Assessment: unknown effectiveness — limited evidence (effectiveness 0%; certainty 25%; harms 0%).*

http://www.conservationevidence.com/actions/765

Remove the chytrid fungus from ponds

One before-and-after study in Mallorca found that drying out a pond and treating resident midwife toads with fungicide reduced levels of infection but did not eradicate chytridiomycosis. *Assessment: unknown effectiveness — limited evidence (effectiveness 25%; certainty 25%; harms 0%).*

http://www.conservationevidence.com/actions/766

Sterilize equipment when moving between amphibian sites

We found no evidence for the effects of sterilizing equipment when moving between amphibian sites on the spread of disease between amphibian populations or individuals. Two randomized, replicated, controlled study in Switzerland and Sweden found that Virkon S disinfectant did not affect survival, mass or behaviour of eggs, tadpoles or hatchlings. However, one of the studies found that bleach significantly reduced tadpole survival. *Assessment: unknown effectiveness — limited evidence (effectiveness 10%; certainty 30%; harms 40%).*

http://www.conservationevidence.com/actions/768

Treating amphibians in the wild or pre-release

One before-and-after study in Mallorca found that treating wild toads with fungicide and drying out the pond reduced infection levels but did not eradicate chytridiomycosis. *Assessment: unknown effectiveness — limited evidence (effectiveness 27%; certainty 30%; harms 0%).*

http://www.conservationevidence.com/actions/767

Use gloves to handle amphibians

We found no evidence for the effects of using gloves on the spread of disease between amphibian populations or individuals. A review for Canada and the USA found that there were no adverse effects of handling 22 amphibian species using disposable gloves. However, three replicated studies in Australia and Austria found that deaths of tadpoles were caused by latex, vinyl and nitrile gloves for 60–100% of species tested. *Assessment: unknown effectiveness — limited evidence (effectiveness 9%; certainty 35%; harms 65%).*

http://www.conservationevidence.com/actions/769

Unlikely to be beneficial

Use antibacterial treatment to reduce infection

Two studies, including one randomized, replicated, controlled study, in New Zealand and Australia found that treatment with chloramphenicol antibiotic, with other interventions in some cases, cured frogs of chytridiomycosis. One replicated, controlled study found that treatment with trimethoprim-sulfadiazine increased survival time but did not cure infected frogs. *Assessment: unlikely to be beneficial (effectiveness 38%; certainty 45%; harms 10%).*

http://www.conservationevidence.com/actions/763

Use antifungal skin bacteria or peptides to reduce infection

Three of four randomized, replicated, controlled studies in the USA found that introducing antifungal bacteria to the skin of chytrid infected amphibians did not reduce infection rate or deaths. One found that it prevented infection and death. One randomized, replicated, controlled study in the USA found that adding antifungal skin bacteria to soil significantly reduced chytridiomycosis infection rate in salamanders. One randomized, replicated, controlled study in Switzerland found that treatment with antimicrobial skin peptides before or after infection with chytridiomycosis did not increase toad survival. *Assessment: unlikely to be beneficial (effectiveness 29%; certainty 50%; harms 10%).*

http://www.conservationevidence.com/actions/764

No evidence found (no assessment)

We have captured no evidence for the following intervention:

- Use zooplankton to remove zoospores

1.8.5 Reduce parasitism and disease – ranaviruses

Based on the collated evidence, what is the current assessment of the effectiveness of interventions for reducing ranaviruses?	
No evidence found (no assessment)	• Sterilize equipment to prevent ranaviruses

No evidence found (no assessment)

We have captured no evidence for the following intervention:

- Sterilize equipment to prevent ranaviruses

1.9 Threat: Pollution

1.9.1 Agricultural pollution

Based on the collated evidence, what is the current assessment of the effectiveness of interventions for agricultural pollution?	
Unknown effectiveness (limited evidence)	• Create walls or barriers to exclude pollutants • Plant riparian buffer strips • Reduce pesticide, herbicide or fertilizer use
No evidence found (no assessment)	• Prevent pollution from agricultural lands or sewage treatment facilities entering watercourses

Unknown effectiveness (limited evidence)

Create walls or barriers to exclude pollutants

One controlled study in Mexico found that installing filters across canals to improve water quality and exclude fish increased weight gain in axolotls. *Assessment: unknown effectiveness — limited evidence (effectiveness 35%; certainty 29%; harms 0%).*

http://www.conservationevidence.com/actions/771

Plant riparian buffer strips

One replicated, controlled study in the USA found that planting buffer strips along streams did not increase amphibian abundance or numbers of species. *Assessment: unknown effectiveness — limited evidence (effectiveness 0%; certainty 30%; harms 0%).*

http://www.conservationevidence.com/actions/819

Reduce pesticide, herbicide or fertilizer use

One study in Taiwan found that halting pesticide use, along with habitat management, increased a population of frogs. *Assessment: unknown effectiveness — limited evidence (effectiveness 71%; certainty 26%; harms 0%).*

http://www.conservationevidence.com/actions/832

No evidence found (no assessment)

We have captured no evidence for the following intervention:

- Prevent pollution from agricultural lands or sewage treatment facilities entering watercourses

1.9.2 Industrial pollution

Based on the collated evidence, what is the current assessment of the effectiveness of interventions for industrial pollution?	
Trade-off between benefit and harms	• Add limestone to water bodies to reduce acidification
No evidence found (no assessment)	• Augment ponds with ground water to reduce acidification

Trade-off between benefit and harms

Add limestone to water bodies to reduce acidification

Five before-and-after studies, including one controlled, replicated study, in the Netherlands and UK found that adding limestone to ponds resulted in establishment of one of three translocated amphibian populations, a temporary increase in breeding and metamorphosis by natterjack toads and increased egg and larval survival of frogs. One replicated, site comparison study in the UK found that habitat management that included adding limestone to ponds increased natterjack toad populations. However, two before-and-after studies, including one controlled study, in the UK found that adding limestone to ponds resulted in increased numbers of abnormal eggs, high tadpole mortality and pond abandonment. *Assessment: trade-offs between benefits and harms (effectiveness 47%; certainty 50%; harms 50%).*

http://www.conservationevidence.com/actions/748

No evidence found (no assessment)

We have captured no evidence for the following intervention:

- Augment ponds with ground water to reduce acidification

1.10 Threat: Climate change and severe weather

Based on the collated evidence, what is the current assessment of the effectiveness of interventions for climate change and severe weather?	
Beneficial	• Deepen ponds to prevent desiccation (deepen, de-silt or re-profile)
Unknown effectiveness (limited evidence)	• Use irrigation systems for amphibian sites (artificially mist habitat)
No evidence found (no assessment)	• Artificially shade ponds to prevent desiccation • Protect habitat along elevational gradients • Provide shelter habitat

Create microclimate and microhabitat refuges

Studies investigating the effects of creating refuges are discussed in 'Habitat restoration and creation' and 'Threat: Biological resource use — Leave coarse woody debris in forests'.

Maintain ephemeral ponds

Studies investigating the effects of regulating water levels and deepening ponds are discussed in 'Threat: Natural system modifications — Regulate water levels' and 'Habitat restoration and creation — Deepen, de-silt or re-profile ponds'.

Beneficial

Deepen ponds to prevent desiccation

Four studies, including one replicated, controlled study, in France, Denmark and the UK found that pond deepening and enlarging or re-profiling resulted in establishment or increased populations of amphibians. Four before-and-after studies in Denmark and the UK found that pond deepening, along with other interventions, maintained newt or increased toad populations. *Assessment for 'Deepen, de-silt or re-profie ponds' from* 'Habitat restoration and creation' section*: beneficial (effectiveness 71%; certainty 65%; harms 0%).*

http://www.conservationevidence.com/actions/806

Unknown effectiveness (limited evidence)

Use irrigation systems for amphibian sites

One before-and-after study in Tanzania found that installing a sprinkler system to mitigate against a reduction of river flow did not maintain a population of Kihansi spray toads. *Assessment for 'Artificially mist habitat to keep it damp' from 'Threat: Energy production and mining' section: unknown effectiveness — limited evidence (effectiveness 24%; certainty 20%; harms 0%).*

http://www.conservationevidence.com/actions/804

No evidence found (no assessment)

We have captured no evidence for the following interventions:

- Artificially shade ponds to prevent desiccation
- Protect habitat along elevational gradients
- Provide shelter habitat

1.11 Habitat protection

Based on the collated evidence, what is the current assessment of the effectiveness of interventions for habitat protection?	
Trade-off between benefit and harms	• Retain buffer zones around core habitat
Unknown effectiveness (limited evidence)	• Protect habitats for amphibians • Retain connectivity between habitat patches

Trade-off between benefit and harms

Retain buffer zones around core habitat

Two studies, including one replicated, controlled study, in Australia and the USA found that retaining unmown buffers around ponds increased numbers of frog species, but had mixed effects on tadpole mass and survival. One replicated, site comparison study in the USA found that retaining buffers along ridge tops within harvested forest increased salamander abundance, body condition and genetic diversity. However, one replicated study in the USA found that 30 m buffer zones around wetlands were not sufficient to protect marbled salamanders. *Assessment: trade-offs between benefits and harms (effectiveness 50%; certainty 50%; harms 25%).*

http://www.conservationevidence.com/actions/850

Unknown effectiveness (limited evidence)

Protect habitats for amphibians

One replicated, site comparison study in the UK found that statutory level habitat protection helped protect natterjack toad populations. One before-and-after study in the UK found that protecting a pond during development had mixed effects on populations of amphibians. *Assessment: unknown effectiveness — limited evidence (effectiveness 51%; certainty 31%; harms 9%).*

http://www.conservationevidence.com/actions/820

Retain connectivity between habitat patches

One before-and-after study in Australia found that retaining native vegetation corridors maintained populations of frogs over 20 years. *Assessment: unknown effectiveness — limited evidence (effectiveness 60%; certainty 31%; harms 0%).*

http://www.conservationevidence.com/actions/853

1.12 Habitat restoration and creation

1.12.1 Terrestrial habitat

Based on the collated evidence, what is the current assessment of the effectiveness of interventions for terrestrial habitat restoration and creation?	
Beneficial	• Replant vegetation
Likely to be beneficial	• Clear vegetation • Create artificial hibernacula or aestivation sites • Create refuges • Restore habitat connectivity
Unknown effectiveness (limited evidence)	• Change mowing regime
No evidence found (no assessment)	• Create habitat connectivity

Beneficial

● Replant vegetation

Four studies, including one replicated study, in Australia, Spain and the USA found that amphibians colonized replanted forest, reseeded grassland and seeded and transplanted upland habitat. Three of four studies, including two replicated studies, in Australia, Canada, Spain and the USA found that areas planted with trees or grass had similar amphibian abundance or community composition to natural sites and one found similar or lower abundance compared to naturally regenerated forest. One found that wetlands within reseeded grasslands were used less than those in natural grasslands. One before-and-after study in Australia found that numbers of frog species increased following restoration that included planting shrubs and trees. *Assessment: beneficial (effectiveness 70%; certainty 63%; harms 3%).*

http://www.conservationevidence.com/actions/849

Likely to be beneficial

Clear vegetation

Seven studies, including four replicated studies, in Australia, Estonia and the UK found that vegetation clearance, along with other habitat management and in some cases release of amphibians, increased or maintained amphibian populations or increased numbers of frog species. However, great crested newt populations were only maintained for six years, but not in the longer term. *Assessment: likely to be beneficial (effectiveness 60%; certainty 54%; harms 10%).*

http://www.conservationevidence.com/actions/761

Create artificial hibernacula or aestivation sites

Two replicated studies in the UK found that artificial hibernacula were used by two of three amphibian species and along with other terrestrial habitat management maintained populations of great crested newts. *Assessment: likely to be beneficial (effectiveness 50%; certainty 44%; harms 0%).*

http://www.conservationevidence.com/actions/759

Create refuges

Two replicated, controlled studies, one of which was randomized, in the USA and Indonesia found that adding coarse woody debris to forest floors had no effect on the number of amphibian species or overall abundance, but had mixed effects on abundance of individual species. One before-and-after study in Australia found that restoration that included reintroducing coarse woody debris to the forest floor increased frog species. Three studies, including two replicated studies, in New Zealand, the UK and USA found that artificial refugia were used by amphibians and, along with other interventions, maintained newt populations. *Assessment: likely to be beneficial (effectiveness 45%; certainty 55%; harms 0%).*

http://www.conservationevidence.com/actions/772

Restore habitat connectivity

One before-and-after study in Italy found that restoring habitat connectivity by raising a road on a viaduct significantly decreased amphibian deaths. *Assessment: likely to be beneficial (effectiveness 75%; certainty 40%; harms 0%).*

http://www.conservationevidence.com/actions/840

Unknown effectiveness (limited evidence)

Change mowing regime

One before-and-after study in Australia found that restoration that included reduced mowing increased numbers of frog species. *Assessment: unknown effectiveness — limited evidence (effectiveness 50%; certainty 30%; harms 0%).*

http://www.conservationevidence.com/actions/783

No evidence found (no assessment)

We have captured no evidence for the following intervention:

- Create habitat connectivity

1.12.2 Aquatic habitat

Based on the collated evidence, what is the current assessment of the effectiveness of interventions for aquatic habitat restoration and creation?	
Beneficial	• Create ponds (amphibians in general) • Create ponds: frogs • Create ponds: natterjack toads • Create ponds: salamanders (including newts) • Create wetlands • Deepen, de-silt or re-profile ponds • Restore wetlands

Likely to be beneficial	• Create ponds: great crested newts • Create ponds: green toads • Create ponds: toads • Remove specific aquatic plants (invasive species) • Restore ponds
Unknown effectiveness (limited evidence)	• Remove tree canopy to reduce pond shading
No evidence found (no assessment)	• Add nutrients to new ponds as larvae food source • Add specific plants to aquatic habitats • Add woody debris to ponds • Create refuge areas in aquatic habitats

Beneficial

Create ponds (amphibians in general)

Twenty-eight studies investigated the colonization of created ponds by amphibians in general, all of which found that amphibians used all or some of the created ponds. Five of nine studies in Australia, Canada, Spain, the UK and USA found that numbers of species were similar or higher in created compared to natural ponds. Nine studies in Europe and the USA found that amphibians established stable populations, used or reproduced in created ponds. Four found that species composition differed, and abundance, juvenile productivity or size in created ponds depended on species. One study found that numbers of species were similar or lower in created ponds. Sixteen studies in Europe and the USA found that created ponds were used or colonized by up to 15 naturally colonizing species, up to 10 species that reproduced or by captive-bred amphibians. Five studies in Europe and the USA found that pond creation, with restoration in three cases, maintained and increased populations or increased species. *Assessment: beneficial (effectiveness 80%; certainty 80%; harms 0%).*

http://www.conservationevidence.com/actions/869

Create ponds (frogs)

Six of nine studies in Australia, Italy, Spain, the UK and USA found that frogs established breeding populations or reproduced in created ponds. One study in Denmark found that frogs colonized created ponds. One study in the Netherlands found that pond creation, along with vegetation clearance, increased frog populations. One study in the USA found that survival increased with age of created ponds. *Assessment: beneficial (effectiveness 75%; certainty 70%; harms 0%).*

http://www.conservationevidence.com/actions/865

Create ponds (natterjack toads)

Five studies in the UK and Denmark found that pond creation, along with other interventions, maintained or increased populations at 75–100% of sites. One study in the UK found that compared to natural ponds, created ponds had lower tadpole mortality from desiccation, but higher mortality from predation by invertebrates. *Assessment: beneficial (effectiveness 75%; certainty 70%; harms 10%).*

http://www.conservationevidence.com/actions/866

Create ponds (salamanders including newts)

Three studies in France, Germany and the USA found that alpine newts, captive-bred smooth newts and translocated spotted salamanders established stable breeding populations in 20–100% of created ponds. Three studies in France, China and the USA found that alpine newts, Chinhai salamanders and translocated spotted salamanders, but not tiger salamanders, reproduced in created ponds. *Assessment: beneficial (effectiveness 70%; certainty 65%; harms 0%).*

http://www.conservationevidence.com/actions/867

Create wetlands

Fifteen studies, including one review and seven replicated studies, in Australia, Kenya and the USA, investigated the effectiveness of creating wetlands for amphibians. Six studies found that created wetlands had similar amphibian abundance, numbers of species or communities as natural wetlands or in one case adjacent forest. Two of those studies found that created wetlands had fewer amphibians, amphibian species and different communities compared to natural wetlands. One global review and two other studies combined created and restored wetlands and found that amphibian abundance and numbers of species were similar or higher compared to natural wetlands. Five of the studies found that up to 15 amphibian species used created wetlands. One study found that captive-bred frogs did not establish in a created wetland. *Assessment: beneficial (effectiveness 75%; certainty 70%; harms 0%).*

http://www.conservationevidence.com/actions/880

Deepen, de-silt or re-profile ponds

Four studies, including one replicated, controlled study, in France, Denmark and the UK found that pond deepening and enlarging or re-profiling resulted in establishment or increased populations of amphibians. Four before-and-after studies in Denmark and the UK found that pond deepening, along with other interventions, maintained newt or increased toad populations. *Assessment: beneficial (effectiveness 71%; certainty 65%; harms 0%).*

http://www.conservationevidence.com/actions/817

Restore wetlands

Seventeen studies, including one review and 11 replicated studies, in Canada, Taiwan and the USA, investigated the effectiveness of wetland restoration for amphibians. Seven of ten studies found that amphibian abundance, numbers of species and species composition were similar in restored and natural wetlands. Two found that abundance or numbers of species were lower and species composition different to natural wetlands. One found mixed results. One global review found that in 89% of cases, restored and created wetlands had similar or higher amphibian abundance or numbers of species to natural wetlands. Seven of nine studies found that wetland restoration increased numbers of amphibian species, with breeding populations establishing in some cases, and maintained or increased abundance of individual species. Three found that amphibian abundance or numbers of species did not increase with restoration. Three of the studies found that restored wetlands were colonized by up to eight amphibian species. *Assessment: beneficial (effectiveness 80%; certainty 73%; harms 0%).*

http://www.conservationevidence.com/actions/879

Likely to be beneficial

Create ponds (great crested newts)

Three studies in Germany and the UK found that great crested newts established breeding populations in created ponds. One systematic review in the UK found that there was no conclusive evidence that mitigation, which often included pond creation, resulted in self-sustaining populations. Four studies in the UK found that great crested newts colonized up to 88% of, or reproduced in 38% of created ponds. *Assessment: likely to be beneficial (effectiveness 60%; certainty 61%; harms 0%).*

http://www.conservationevidence.com/actions/863

Create ponds (green toads)

Two studies in Denmark found that pond creation, along with other interventions, significantly increased green toad populations. One study in Sweden found that green toads used or reproduced in 41–59% of created ponds. *Assessment: likely to be beneficial (effectiveness 73%; certainty 59%; harms 0%).*

http://www.conservationevidence.com/actions/864

Create ponds (toads)

Five studies in Germany, Switzerland, the UK and USA found that toads established breeding populations or reproduced in 16–100% of created ponds. Two studies in Denmark and Switzerland found that wild but not captive-bred toads colonized 29–100% of created ponds. One study in Denmark found that creating ponds, along with other interventions, increased toad populations. *Assessments: likely to be beneficial (effectiveness 70%; certainty 60%; harms 0%).*

http://www.conservationevidence.com/actions/868

Remove specific aquatic plants

One before-and-after study in the UK found that habitat and species management that included controlling swamp stonecrop, increased a population of natterjack toads. One replicated, controlled study in the USA found that more Oregon spotted frogs laid eggs in areas where invasive reed canarygrass was mown. *Assessment for 'Control invasive plants' from* 'Threat: Invasive alien and other problematic species': *likely to be beneficial (effectiveness 60%; certainty 47%; harms 0%).*

http://www.conservationevidence.com/actions/815

Restore ponds

Fifteen studies investigated the effectiveness of pond restoration for amphibians. Three studies, including one replicated, controlled, before-and-after study in Denmark, the UK and USA found that pond restoration did not increase or had mixed effects on population numbers and hatching success. One replicated, before-and-after study in the UK found that restoration increased pond use. One replicated study in Sweden found that only 10% of restored ponds were used for breeding. Three before-and-after studies, including one replicated, controlled study, in Denmark and Italy found that restored and created ponds were colonized by up to seven species. Eight of nine studies, including one systematic review, in Denmark, Estonia, Italy and the UK found that pond restoration, along with other habitat management, maintained or increased populations, or increased pond occupancy, ponds with breeding success or numbers of amphibian species. One found that numbers of species did not increase and one found that great crested newt populations did not establish. *Assessment: likely to be beneficial (effectiveness 60%; certainty 63%; harms 0%).*

http://www.conservationevidence.com/actions/878

Unknown effectiveness (limited evidence)

Remove tree canopy to reduce pond shading

One before-and-after study in the USA found that canopy removal did not increase hatching success of spotted salamanders. One before-and-after study in Denmark found that following pond restoration that included canopy removal, translocated toads established breeding populations. *Assessment: unknown effectiveness — limited evidence (effectiveness 30%; certainty 25%; harms 0%).*

http://www.conservationevidence.com/actions/758

No evidence found (no assessment)

We have captured no evidence for the following interventions:

- Add nutrients to new ponds as larvae food source
- Add specific plants to aquatic habitats
- Add woody debris to ponds
- Create refuge areas in aquatic habitats

1.13 Species management

Strict protocols should be followed when carrying out these interventions to minimise potential spread of disease-causing agents such as chytrid fungi and Ranavirus.

1.13.1 Translocate amphibians

Based on the collated evidence, what is the current assessment of the effectiveness of translocations?	
Likely to be beneficial	• Translocate amphibians (amphibians in general) • Translocate amphibians (great crested newts) • Translocate amphibians (natterjack toads) • Translocate amphibians (salamanders including newts) • Translocate amphibians (toads) • Translocate amphibians (wood frogs)
Trade-off between benefit and harms	• Translocate amphibians (frogs)

Likely to be beneficial

Translocate amphibians (amphibians in general)

Overall, three global reviews and one study in the USA found that 65% of amphibian translocations that could be assessed resulted in established breeding populations or substantial recruitment to the adult population. A further two translocations resulted in breeding and one in survival following release. One review found that translocations of over 1,000 animals were more successful, but that success was not related to the source of animals (wild or captive), life-stage, continent or reason for translocation. *Assessment: likely to be beneficial (effectiveness 60%; certainty 60%; harms 19%).*

http://www.conservationevidence.com/actions/854

Translocate amphibians (great crested newts)

Four of six studies in the UK found that translocated great crested newts maintained or established breeding populations. One found that populations survived at least one year in 37% of cases, but one found that within three years breeding failed in 48% of ponds. A systematic review of 31 studies found no conclusive evidence that mitigation that included translocations resulted in self-sustaining populations. One review found that newts reproduced following 56% of translocations, in some cases along with other interventions. *Assessment: likely to be beneficial (effectiveness 50%; certainty 50%; harms 10%).*

http://www.conservationevidence.com/actions/858

Translocate amphibians (natterjack toads)

Three studies in France and the UK found that translocated natterjack toad eggs, tadpoles, juveniles or adults established breeding populations at some sites, although head-started or captive-bred animals were also released at some sites. Re-establishing toads on dune or saltmarsh habitat was more successful than on heathland. One study in the UK found that repeated translocations of wild rather than captive-bred toads were more successful. *Assessment: likely to be beneficial (effectiveness 60%; certainty 56%; harms 10%).*

http://www.conservationevidence.com/actions/859

Translocate amphibians (salamanders including newts)

Four studies in the UK and USA found that translocated eggs or adults established breeding populations of salamanders or smooth newts. One study in the USA found that one of two salamander species reproduced following translocation of eggs, tadpoles and metamorphs. One study in the USA found that translocated salamander eggs hatched and tadpoles had similar survival rates as in donor ponds. *Assessment: likely to be beneficial (effectiveness 70%; certainty 55%; harms 0%).*

http://www.conservationevidence.com/actions/860

Translocate amphibians (toads)

Two of four studies in Denmark, Germany, the UK and USA found that translocating eggs and/or adults established common toad breeding populations. One found populations of garlic toads established at two of four sites and one that breeding populations of boreal toads were not established. One study in Denmark found that translocating green toad eggs to existing populations, along with habitat management, increased population numbers. Four studies in Germany, Italy, South Africa and the USA found that translocated adult toads reproduced, survived up to six or 23 years, or some metamorphs survived over winter. *Assessment: likely to be beneficial (effectiveness 60%; certainty 56%; harms 10%).*

http://www.conservationevidence.com/actions/855

Translocate amphibians (wood frogs)

Two studies in the USA found that following translocation of wood frog eggs, breeding populations were established in 25–50% of created ponds. One study in the USA found that translocated eggs hatched and up to 57% survived as tadpoles in pond enclosures. *Assessment: likely to be beneficial (effectiveness 40%; certainty 50%; harms 0%).*

http://www.conservationevidence.com/actions/856

Trade-off between benefit and harms

Translocate amphibians (frogs)

Eight of ten studies in New Zealand, Spain, Sweden, the UK and USA found that translocating frog eggs, juveniles or adults established breeding populations. Two found that breeding populations went extinct within five years or did not establish. Five studies in Canada, New Zealand and the USA found that translocations of eggs, juveniles or adults resulted in little or no breeding at some sites. Five studies in Italy, New Zealand and the USA found that translocated juveniles or adults survived the winter or up to eight years. One study in the USA found that survival was lower for Oregon spotted frogs translocated as adults compared to eggs. Two studies in the USA found that 60–100% of translocated frogs left the release site and 35–73% returned to their original pond within 32 days. Two studies in found that frogs either lost or gained weight after translocation. *Assessment: trade-offs between benefits and harms (effectiveness 58%; certainty 65%; harms 20%).*

http://www.conservationevidence.com/actions/861

1.13.2 Captive breeding, rearing and releases

Based on the collated evidence, what is the current assessment of the effectiveness of captive breeding, rearing and releases?	
Likely to be beneficial	• Release captive-bred individuals (amphibians in general) • Release captive-bred individuals: frogs
Trade-off between benefit and harms	• Breed amphibians in captivity: frogs • Breed amphibians in captivity: harlequin toads • Breed amphibians in captivity: Mallorcan midwife toad • Breed amphibians in captivity: salamanders (including newts) • Breed amphibians in captivity: toads • Head-start amphibians for release • Release captive-bred individuals: Mallorcan midwife toads • Release captive-bred individuals: toads • Use artificial fertilization in captive breeding • Use hormone treatment to induce sperm and egg release
Unknown effectiveness (limited evidence)	• Release captive-bred individuals: salamanders (including newts)
Unlikely to be beneficial	• Freeze sperm or eggs for future use
Likely to be ineffective or harmful	• Release captive-bred individuals: green and golden bell frogs

Likely to be beneficial

Release captive-bred individuals (amphibians in general)

One review found that 41% of release programmes of captive-bred or head-started amphibians showed evidence of breeding in the wild for multiple generations, 29% showed some evidence of breeding and 12% evidence of survival following release. *Assessment: likely to be beneficial (effectiveness 55%; certainty 50%; harms 10%).*

http://www.conservationevidence.com/actions/871

Release captive-bred individuals (frogs)

Five of six studies in Europe, Hong Kong and the USA found that captive-bred frogs released as tadpoles, juveniles or adults established breeding populations and in some cases colonized new sites. Three studies in Australia and the USA found that a high proportion of frogs released as eggs survived to metamorphosis, some released tadpoles survived the first few months, but few released froglets survived. Four studies in Australia, Italy, the UK and USA found that captive-bred frogs reproduced at 31–100% of release sites, or that breeding was limited. *Assessment: likely to be beneficial (effectiveness 60%; certainty 60%; harms 15%).*

http://www.conservationevidence.com/actions/870

Trade-off between benefit and harms

Breed amphibians in captivity (frogs)

Twenty-three of 33 studies across the world found that amphibians produced eggs in captivity. Seven found mixed results, with some species or populations reproducing successfully, but with other species difficult to maintain or raise to adults. Two found that frogs did not breed successfully or died in captivity. Seventeen of the studies found that captive-bred frogs were raised successfully to hatching, tadpoles, froglets or adults in captivity. Four studies in Canada, Fiji, Hong Kong and Italy found that 30–88% of eggs hatched, or survival to metamorphosis was 75%, as froglets was 17–51% or to adults was 50–90%. *Assessment: trade-offs between benefits and harms (effectiveness 60%; certainty 68%; harms 30%).*

http://www.conservationevidence.com/actions/835

Breed amphibians in captivity (harlequin toads)

Four of five studies in Colombia, Ecuador, Germany and the USA found that harlequin toads reproduced in captivity. One found that eggs were only produced by simulating a dry and wet season and one found that breeding was difficult. One found that captive-bred harlequin toads were raised successfully to metamorphosis in captivity and two found that most toads died before or after hatching. *Assessment: trade-offs between benefits and harms (effectiveness 44%; certainty 50%; harms 28%).*

http://www.conservationevidence.com/actions/836

Breed amphibians in captivity (Mallorcan midwife toad)

Two studies in the UK found that Mallorcan midwife toads produced eggs that were raised to metamorphs or toadlets in captivity. However, clutches dropped by males were not successfully maintained artificially. One study in the UK found that toads bred in captivity for nine or more generations had slower development, reduced genetic diversity and predator defence traits. *Assessment: trade-offs between benefits and harms (effectiveness 69%; certainty 55%; harms 40%).*

http://www.conservationevidence.com/actions/837

Breed amphibians in captivity (salamanders including newts)

Four of six studies in Japan, Germany, the UK and USA found that eggs were produced successfully in captivity. Captive-bred salamanders were raised to yearlings, larvae or adults. One review found that four of five salamander species bred successfully in captivity. Four studies in Germany, Mexico and the USA found that egg production, larval development, body condition and survival were affected by water temperature, density or enclosure type. *Assessment: trade-offs between benefits and harms (effectiveness 60%; certainty 50%; harms 25%).*

http://www.conservationevidence.com/actions/838

Breed amphibians in captivity (toads)

Ten studies in Germany, Italy, Spain, the UK and USA found that toads produced eggs in captivity. Eight found that toads were raised successfully to tadpoles, toadlets or adults in captivity. Two found that most died after hatching or metamorphosis. Two reviews found mixed results with four species of toad or 21% of captive populations of Puerto Rican crested toads breeding successfully. Four studies in Germany, Spain and the USA found that reproductive success was affected by tank location and humidity. *Assessment: trade-offs between benefits and harms (effectiveness 65%; certainty 60%; harms 25%).*

http://www.conservationevidence.com/actions/848

Head-start amphibians for release

Twenty-two studies head-started amphibians from eggs and monitored them after release. A global review and six of 10 studies in Europe and the USA found that released head-started tadpoles, metamorphs or juveniles established breeding populations or increased existing populations. Two found mixed results with breeding populations established in 71% of studies reviewed or at 50% of sites. Two found that head-started metamorphs or adults did not establish a breeding population or prevent a population decline. An additional 10 studies in Australia, Canada, Europe and the USA measured aspects of survival or breeding success of released head-started amphibians and found mixed results. Three studies in the USA only provided results for head-starting in captivity. Two of those found that eggs could be reared to tadpoles, but only one successfully reared adults. *Assessment: trade-offs between benefits and harms (effectiveness 60%; certainty 60%; harms 25%).*

http://www.conservationevidence.com/actions/881

Release captive-bred individuals (Mallorcan midwife toad)

Three studies in Mallorca found that captive-bred midwife toads released as tadpoles, toadlets or adults established breeding populations at 38–100% of sites. One study in the UK found that predator defences were maintained, but genetic diversity was reduced in a captive-bred population. *Assessment: trade-offs between benefits and harms (effectiveness 68%; certainty 58%; harms 20%).*

http://www.conservationevidence.com/actions/873

Release captive-bred individuals (toads)

Two of three studies in Denmark, Sweden and the USA found that captive-bred toads released as tadpoles, juveniles or metamorphs established populations. The other found that populations were not established. Two studies in Puerto Rico found that survival of released captive-bred Puerto Rican crested toads was low. *Assessment: trade-offs between benefits and harms (effectiveness 40%; certainty 50%; harms 20%).*

http://www.conservationevidence.com/actions/875

Use artificial fertilization in captive breeding

Three replicated studies, including two randomized studies, in Australia and the USA found that the success of artificial fertilization depended on the type and number of doses of hormones used to stimulate egg production. One replicated study in Australia found that 55% of eggs were fertilized artificially, but soon died. *Assessment: trade-offs between benefits and harms (effectiveness 40%; certainty 40%; harms 20%).*

http://www.conservationevidence.com/actions/834

Use hormone treatment to induce sperm and egg release

One review and nine of 10 replicated studies, including two randomized, controlled studies, in Austria, Australia, China, Latvia, Russia and the USA found that hormone treatment of male amphibians stimulated or increased sperm production, or resulted in successful breeding. One found that hormone treatment of males and females did not result in breeding. One review and nine of 14 replicated studies, including six randomized and/or controlled studies, in Australia, Canada, China, Ecuador, Latvia and the USA found that hormone treatment of female amphibians had mixed results, with 30–71% of females producing viable eggs following treatment, or with egg production depending on the combination, amount or number of doses of hormones. Three found that hormone treatment stimulated egg production or successful breeding. Two found that treatment did not stimulate or increase egg production. *Assessment: trade-offs between benefits and harms (effectiveness 50%; certainty 65%; harms 30%).*

http://www.conservationevidence.com/actions/883

Unknown effectiveness (limited evidence)

Release captive-bred individuals (salamanders including newts)

One study in Germany found that captive-bred great crested newts and smooth newts released as larvae, juveniles and adults established stable breeding populations. *Assessment: unknown effectiveness — limited evidence (effectiveness 70%; certainty 30%; harms 0%).*

http://www.conservationevidence.com/actions/874

Unlikely to be beneficial

Freeze sperm or eggs for future use

Ten replicated studies, including three controlled studies, in Austria, Australia, Russia, the UK and USA found that following freezing, viability of amphibian sperm, and in one case eggs, depended on species, cryoprotectant used, storage temperature or method and freezing or thawing rate. One found that sperm could be frozen for up to 58 weeks. *Assessment: unlikely to be beneficial (effectiveness 35%; certainty 50%; harms 10%).*

http://www.conservationevidence.com/actions/876

Likely to be ineffective or harmful

Release captive-bred individuals (green and golden bell frogs)

Three studies in Australia found that captive-bred green and golden bell frogs released mainly as tadpoles did not established breeding populations, or only established breeding populations in 25% of release programmes. One study in Australia found that some frogs released as tadpoles survived at least 13 months. *Assessment: likely to be ineffective or harmful (effectiveness 20%; certainty 50%; harms 20%).*

http://www.conservationevidence.com/actions/872

1.14 Education and awareness raising

Based on the collated evidence, what is the current assessment of the effectiveness of interventions for education and awareness raising?	
Likely to be beneficial	• Engage volunteers to collect amphibian data (citizen science) • Provide education programmes about amphibians • Raise awareness amongst the general public through campaigns and public information

Likely to be beneficial

Engage volunteers to collect amphibian data (citizen science)

Five studies in Canada, the UK and USA found that amphibian data collection projects engaged up to 10,506 volunteers and were active in 16–17 states in the USA. Five studies in the UK and USA found that volunteers surveyed up to 7,872 sites, swabbed almost 6,000 amphibians and submitted thousands of amphibian records. *Assessment: likely to be beneficial (effectiveness 66%; certainty 60%; harms 0%).*

http://www.conservationevidence.com/actions/760

Provide education programmes about amphibians

One study in Taiwan found that education programmes about wetlands and amphibians, along with other interventions, doubled a population of Taipei frogs. Four studies, including one replicated study, in Germany, Mexico,

Slovenia, Zimbabwe and the USA found that education programmes increased the amphibian knowledge of students. *Assessment: likely to be beneficial (effectiveness 58%; certainty 55%; harms 0%).*

http://www.conservationevidence.com/actions/776

Raise awareness amongst the general public through campaigns and public information

Two studies, including one replicated, before-and-after study, in Estonia and the UK found that raising public awareness, along with other interventions, increased amphibian breeding habitat and numbers of toads. One before-and-after study in Mexico found that raising awareness in tourists increased their knowledge of axolotls. However, one study in Taiwan found that holding press conferences had no effect on a frog conservation project. *Assessment: likely to be beneficial (effectiveness 60%; certainty 51%; harms 0%).*

http://www.conservationevidence.com/actions/831

2. BAT CONSERVATION

Anna Berthinussen, Olivia C. Richardson, Rebecca K. Smith, John D. Altringham & William J. Sutherland

Expert assessors

John Altringham, University of Leeds, UK
James Aegerter, Animal and Plant Health Agency, UK
Kate Barlow, Bat Conservation Trust, UK
Anna Berthinussen, University of Leeds, UK
Fabio Bontadina, SWILD — Urban Ecology & Wildlife Research, Switzerland
David Bullock, National Trust, UK
Paul Cryan, Fort Collins Science Center, US Geological Survey
Brock Fenton, University of Western Ontario, Canada
Anita Glover, University of Leeds, UK
Joanne Hodgkins, National Trust, UK
David Jacobs, University of Cape Town, South Africa
Bradley Law, Forest Science Centre, New South Wales Department of Primary Industries, Australia
Christoph Meyer, Centre for Environmental Biology, Portugal
Kirsty Park, University of Stirling, UK
Guido Reiter, Co-ordination Centre for Bat Conservation and Research in Austria
Danilo Russo, Federico II University of Naples, Italy
Rebecca K. Smith, University of Cambridge, UK
Matt Struebig, University of Kent, UK
Christian Voigt, Freie Universität Berlin, Germany
Michael Willig, University of Connecticut, USA

Scope of assessment: for native wild bat species across the world.

Assessed: 2015.

Effectiveness measure is the median % score for effectiveness.

Certainty measure is the median % certainty of evidence for effectiveness, determined by the quantity and quality of the evidence in the synopsis.

Harm measure is the median % score for negative side-effects to the group of species of concern.

 http://dx.doi.org/10.11647/OBP.0109.02

This book is meant as a guide to the evidence available for different conservation interventions and as a starting point in assessing their effectiveness. The assessments are based on the available evidence for the target group of species for each intervention. The assessment may therefore refer to different species or habitat to the one(s) you are considering. Before making any decisions about implementing interventions it is vital that you read the more detailed accounts of the evidence in order to assess their relevance for your study species or system.

Full details of the evidence are available at
www.conservationevidence.com

There may also be significant negative side-effects on the target groups or other species or communities that have not been identified in this assessment.

A lack of evidence means that we have been unable to assess whether or not an intervention is effective or has any harmful impacts.

2.1 Threat: Residential and commercial development

Based on the collated evidence, what is the current assessment of the effectiveness of interventions for residential and commercial development?	
Unknown effectiveness (limited evidence)	• Protect brownfield sites • Provide foraging habitat in urban areas
No evidence found (no assessment)	• Change timing of building works • Conserve existing roosts within developments • Conserve old buildings or structures as roosting sites for bats within developments • Create alternative roosts within buildings • Maintain bridges and retain crevices for roosting • Retain or relocate access points to bat roosts • Retain or replace existing bat commuting routes within development

Unknown effectiveness (limited evidence)

Protect brownfield sites

One study in the USA found bat activity within an urban wildlife refuge on an abandoned manufacturing site to be consistent with predictions across North America based on the availability of potential roosts. *Assessment: unknown effectiveness (effectiveness 40%; certainty 20%; harms 0%).*

http://www.conservationevidence.com/actions/953

Provide foraging habitat in urban areas

One site comparison study in the USA found higher bat activity in restored forest preserves in urban areas than in an unrestored forest preserve. One replicated, controlled, site comparison study in the UK found higher bat activity over green roofs in urban areas than conventional unvegetated roofs. *Assessment: unknown effectiveness (effectiveness 50%; certainty 30%; harms 0%).*

http://www.conservationevidence.com/actions/954

No evidence found (no assessment)

We have captured no evidence for the following interventions:

- Change timing of building works
- Conserve existing roosts within developments
- Conserve old buildings or structures as roosting sites for bats within developments
- Create alternative roosts within buildings
- Maintain bridges and retain crevices for roosting
- Retain or relocate access points to bat roosts
- Retain or replace existing bat commuting routes within development

2.2 Threat: Agriculture

2.2.1 Land use change

Based on the collated evidence, what is the current assessment of the effectiveness of interventions for land use change?	
Likely to be beneficial	• Protect or create wetlands as foraging habitat for bats
Unknown effectiveness (limited evidence)	• Retain or plant trees on agricultural land to replace foraging habitat for bats
No evidence found (no assessment)	• Conserve old buildings or structures on agricultural land as roosting sites for bats • Retain old or dead trees with hollows and cracks as roosting sites for bats on agricultural land • Retain or replace existing bat commuting routes on agricultural land

Likely to be beneficial

Protect or create wetlands as foraging habitat for bats

We found no evidence for the effects of protecting existing wetlands. One replicated, controlled, site comparison study in the USA found higher bat activity over heliponds and drainage ditches within a pine plantation than over natural wetlands. A replicated study in Germany found high levels of bat activity over constructed retention ponds compared to nearby vineyard sites, but comparisons were not made with natural pond sites. *Assessment: likely to be beneficial (effectiveness 60%; certainty 48%; harms 0%).*

http://www.conservationevidence.com/actions/959

Unknown effectiveness (limited evidence)

Retain or plant trees on agricultural land to replace foraging habitat

We found no evidence for the effects of retaining trees as foraging habitat for bats. Two site comparison studies (one replicated) in Australia found no difference in bat activity and the number of bat species in agricultural areas revegetated with native plantings and over grazing land without trees. In both studies, bat activity was lower in plantings than in original forest and woodland remnants. *Assessment: unknown effectiveness (effectiveness 20%; certainty 20%; harms 0%).*

http://www.conservationevidence.com/actions/958

No evidence found (no assessment)

We have captured no evidence for the following interventions:

- Conserve old buildings or structures on agricultural land as roosting sites for bats
- Retain old or dead trees with hollows and cracks as roosting sites for bats on agricultural land
- Retain or replace existing bat commuting routes on agricultural land

2.2.2 Intensive farming

Based on the collated evidence, what is the current assessment of the effectiveness of interventions for intensive farming?	
Likely to be beneficial	• Convert to organic farming • Encourage agroforestry
Unknown effectiveness (limited evidence)	• Introduce agri-environment schemes

Likely to be beneficial

Convert to organic farming

Four replicated, paired, site comparison studies on farms in the UK had inconsistent results. Two studies found higher bat abundance and activity on organic farms than conventional farms, and two studies showed no difference in bat abundance between organic and non-organic farms. *Assessment: likely to be beneficial (effectiveness 40%; certainty 40%; harms 0%).*

http://www.conservationevidence.com/actions/961

Encourage agroforestry

Four replicated, site comparison studies (three in Mexico and one in Costa Rica) found no difference in bat diversity, the number of bat species and/or bat abundance between cacao, coffee or banana agroforestry plantations and native rainforest. One replicated, site comparison study in Mexico found higher bat diversity in native forest fragments than in coffee agroforestry plantations. One replicated, randomized, site comparison study in Costa Rica found lower bat diversity in native rainforest than in cacao agroforestry plantations. A replicated, site comparison study in Mexico found that bat diversity in coffee agroforestry plantations and native rainforest was affected by the proportion of each habitat type within the landscape. Three studies found that increasing management intensity on agroforestry plantations had a negative effect on some bat species, and a positive effect on others. *Assessment: likely to be beneficial (effectiveness 50%; certainty 50%; harms 10%).*

http://www.conservationevidence.com/actions/963

Unknown effectiveness (limited evidence)

Introduce agri-environment schemes

One replicated, paired study in Scotland, UK found lower bat activity on farms participating in agri-environment schemes than on non-participating conventional farms. *Assessment: unknown effectiveness (effectiveness 0%; certainty 18%; harms 13%).*

http://www.conservationevidence.com/actions/962

2.3 Threat: Energy production – wind turbines

Based on the collated evidence, what is the current assessment of the effectiveness of interventions for wind turbines?	
Beneficial	• Switch off turbines at low wind speeds to reduce bat fatalities
Likely to be beneficial	• Deter bats from turbines using ultrasound
Unknown effectiveness (limited evidence)	• Deter bats from turbines using radar
No evidence found (no assessment)	• Automatically switch off wind turbines when bat activity is high • Close off nacelles on wind turbines to prevent roosting bats • Leave a minimum distance between turbines and habitat features used by bats • Modify turbine design to reduce bat fatalities • Modify turbine placement to reduce bat fatalities • Remove turbine lighting to avoid attracting bats

Beneficial

● Switch off turbines at low wind speeds to reduce bat fatalities

Three replicated, controlled studies in Canada and the USA have shown that reducing the operation of wind turbines at low wind speeds causes a reduction in bat fatalities. *Assessment: beneficial (effectiveness 80%; certainty 70%; harms 0%).*

http://www.conservationevidence.com/actions/970

Likely to be beneficial

Deter bats from turbines using ultrasound

Five field studies at wind farms or pond sites (including one replicated, randomized, before-and-after trial), and one laboratory study, have all found lower bat activity or fewer bat deaths with ultrasonic deterrents than without. *Assessment: likely to be beneficial (effectiveness 50%; certainty 40%; harms 10%).*

http://www.conservationevidence.com/actions/968

Unknown effectiveness (limited evidence)

Deter bats from turbines using radar

A replicated, site comparison study in the UK found reduced bat activity in natural habitats in proximity to electromagnetic fields produced by radars. We found no evidence for the effects of installing radars on wind turbines on bats. *Assessment: unknown effectiveness (effectiveness 20%; certainty 10%; harms 0%).*

http://www.conservationevidence.com/actions/967

No evidence found (no assessment)

We have captured no evidence for the following interventions:

- Automatically switch off wind turbines when bat activity is high
- Close off nacelles on wind turbines to prevent roosting bats
- Leave a minimum distance between turbines and habitat features used by bats
- Modify turbine design to reduce bat fatalities
- Modify turbine placement to reduce bat fatalities
- Remove turbine lighting to avoid attracting bats

2.4 Threat: Energy production – mining

Based on the collated evidence, what is the current assessment of the effectiveness of interventions for mining?	
No evidence found (no assessment)	• Legally protect bat hibernation sites in mines from reclamation • Provide artificial hibernacula to replace roosts lost in reclaimed mines • Relocate bats from reclaimed mines to new hibernation sites

No evidence found (no assessment)

We have captured no evidence for the following interventions:

- Legally protect bat hibernation sites in mines from reclamation
- Provide artificial hibernacula to replace roosts lost in reclaimed mines
- Relocate bats from reclaimed mines to new hibernation sites

2.5 Threat: Transportation and service corridors

Based on the collated evidence, what is the current assessment of the effectiveness of interventions for roads?	
Likely to be beneficial	• Install underpasses as road crossing structures for bats
Unknown effectiveness (limited evidence)	• Divert bats to safe crossing points with plantings or fencing • Install bat gantries or bat bridges as road crossing structures for bats • Install overpasses as road crossing structures for bats
No evidence found (no assessment)	• Deter bats with lighting • Install green bridges as road crossing structures for bats • Install hop-overs as road crossing structures for bats • Replace or improve habitat for bats around roads

Likely to be beneficial

Install underpasses as road crossing structures for bats

Four studies (two replicated) in Germany, Ireland and the UK found varying proportions of bats to be using existing underpasses below roads and crossing over the road above. *Assessment: likely to be beneficial (effectiveness 60%; certainty 50%; harms 0%).*

http://www.conservationevidence.com/actions/976

Unknown effectiveness (limited evidence)

Divert bats to safe crossing points with plantings or fencing

We found no evidence for the effects of diverting bats to safe road crossing points. One controlled, before-and-after study in Switzerland found that a small proportion of lesser horseshoe bats within a colony flew along an artificial hedgerow to commute. *Assessment: unknown effectiveness (effectiveness 10%; certainty 10%; harms 5%).*

http://www.conservationevidence.com/actions/981

Install bat gantries or bat bridges as road crossing structures for bats

One replicated, site comparison study in the UK found fewer bats using bat gantries than crossing the road below at traffic height. *Assessment: unknown effectiveness (effectiveness 0%; certainty 35%; harms 0%).*

http://www.conservationevidence.com/actions/978

Install overpasses as road crossing structures for bats

One replicated, site comparison study in Ireland did not find more bats using over-motorway routes than crossing over the road below. *Assessment: unknown effectiveness (effectiveness 0%; certainty 20%; harms 0%).*

http://www.conservationevidence.com/actions/977

No evidence found (no assessment)

We have captured no evidence for the following interventions:

- Deter bats with lighting
- Install green bridges as road crossing structures for bats
- Install hop-overs as road crossing structures for bats
- Replace or improve habitat for bats around roads

2.6 Threat: Biological resource use

2.6.1 Hunting

Based on the collated evidence, what is the current assessment of the effectiveness of interventions for hunting?	
No evidence found (no assessment)	• Educate local communities about bats and hunting • Introduce and enforce legislation to control hunting of bats • Introduce sustainable harvesting of bats

No evidence found (no assessment)

We have captured no evidence for the following interventions:

- Educate local communities about bats and hunting
- Introduce and enforce legislation to control hunting of bats
- Introduce sustainable harvesting of bats

2.6.2 Guano harvesting

Based on the collated evidence, what is the current assessment of the effectiveness of interventions for guano harvesting?	
No evidence found (no assessment)	• Introduce and enforce legislation to regulate the harvesting of bat guano • Introduce sustainable harvesting of bat guano

No evidence found (no assessment)

We have captured no evidence for the following interventions:

- Introduce and enforce legislation to regulate the harvesting of bat guano
- Introduce sustainable harvesting of bat guano

2.6.3 Logging and wood harvesting

Based on the collated evidence, what is the current assessment of the effectiveness of interventions for logging and wood harvesting?	
Likely to be beneficial	• Incorporate forested corridors or buffers into logged areas • Use selective harvesting/reduced impact logging instead of clearcutting • Use shelterwood cutting instead of clearcutting
Unknown effectiveness (limited evidence)	• Retain residual tree patches in logged areas • Thin trees within forests
No evidence found (no assessment)	• Manage woodland or forest edges for bats • Replant native trees • Retain deadwood/snags within forests for roosting bats

Likely to be beneficial

Incorporate forested corridors or buffers into logged areas

One replicated, site comparison study in Australia found no difference in the activity and number of bat species between riparian buffers in logged, regrowth or mature forest. One replicated, site comparison study in North America found higher bat activity along the edges of forested corridors than in corridor interiors or adjacent logged stands. Three replicated, site comparison studies in Australia and North America found four bat species roosting in forested corridors and riparian buffers. *Assessment: likely to be beneficial (effectiveness 50%; certainty 50%; harms 0%).*

http://www.conservationevidence.com/actions/996

Use selective harvesting/reduced impact logging instead of clearcutting

Nine replicated, controlled, site comparison studies provide evidence for the effects of selective or reduced impact logging on bats with mixed results. One study in the USA found that bat activity was higher in selectively logged forest than in unharvested forest. One study in Italy caught fewer barbastelle bats in selectively logged forest than in unmanaged forest. Three studies in Brazil and two in Trinidad found no difference in bat abundance or species diversity between undisturbed control forest and selectively logged or reduced impact logged forest, but found differences in species composition. Two studies in Brazil found no effect of reduced impact logging on the activity of the majority of bat species, but mixed effects on the activity of four species. *Assessment: likely to be beneficial (effectiveness 60%; certainty 50%; harms 10%).*

http://www.conservationevidence.com/actions/989

Use shelterwood cutting instead of clearcutting

One site comparison study in North America found higher or equal activity of at least five bat species in shelterwood harvests compared to unharvested control sites. One replicated, site comparison study in Australia found Gould's long eared bats selectively roosting in shelterwood harvests, but southern forest bats roosting more often in mature unlogged forest. A replicated, site comparison study in Italy found barbastelle bats favoured unmanaged woodland for roosting and used shelterwood harvested woodland in proportion to availability. *Assessment: likely to be beneficial (effectiveness 50%; certainty 48%; harms 18%).*

http://www.conservationevidence.com/actions/990

Unknown effectiveness (limited evidence)

Retain residual tree patches in logged areas

Two replicated, site comparison studies in Canada found no difference in bat activity between residual tree patch edges in clearcut blocks and edges of the remaining forest. One of the studies found higher activity of smaller bat species at residual tree patch edges than in the centre of open clearcut blocks. Bat activity was not compared to unlogged areas. *Assessment: unknown effectiveness (effectiveness 20%; certainty 25%; harms 0%).*

http://www.conservationevidence.com/actions/995

Thin trees within forests

Two replicated, site comparison studies (one paired) in North America found that bat activity was higher in thinned forest stands than in unthinned stands, and similar to that in mature forest. One replicated, site comparison study in North America found higher bat activity in thinned than in unthinned forest stands in one of the two years of the study. One replicated, site comparison study in Canada found the silver-haired bat more often in clearcut patches than unthinned forest, but found no difference in the activity of Myotis species. *Assessment: unknown effectiveness (effectiveness 45%; certainty 38%; harms 10%).*

http://www.conservationevidence.com/actions/991

No evidence found (no assessment)

We have captured no evidence for the following interventions:

- Manage woodland or forest edges for bats
- Replant native trees
- Retain deadwood/snags within forests for roosting bats

2.7 Threat: Human disturbance – caving and tourism

Based on the collated evidence, what is the current assessment of the effectiveness of interventions for caving and tourism?	
Likely to be beneficial	• Impose restrictions on cave visits
Trade-offs between benefit and harms	• Use cave gates to restrict public access
No evidence found (no assessment)	• Educate the public to reduce disturbance to hibernating bats • Legally protect bat hibernation sites • Maintain microclimate at underground hibernation/roost sites • Provide artificial hibernacula for bats to replace disturbed sites

Likely to be beneficial

Impose restrictions on cave visits

Two before-and-after studies from Canada and Turkey found that bat populations within caves increased after restrictions on cave visitors were imposed. *Assessment: likely to be beneficial (effectiveness 70%; certainty 50%; harms 0%).*

http://www.conservationevidence.com/actions/1002

Trade-off between benefit and harms

Use cave gates to restrict public access

Ten studies in Europe, North America and Australia provide evidence for the effects of cave gating on bats, with mixed results. Four of the studies (one replicated) found more or equal numbers of bats in underground systems after gating. Two of the studies (one replicated) found reduced bat populations or incidences of cave abandonment after gating. Five studies (two replicated) provide evidence for changes in flight behaviour at cave gates. *Assessment: trade-offs between benefits and harms (effectiveness 50%; certainty 60%; harms 50%).*

http://www.conservationevidence.com/actions/999

No evidence found (no assessment)

We have captured no evidence for the following interventions:

- Educate the public to reduce disturbance to hibernating bats
- Legally protect bat hibernation sites
- Maintain microclimate at underground hibernation/roost sites
- Provide artificial hibernacula for bats to replace disturbed sites

2.8 Threat: Natural system modification – natural fire and fire suppression

Based on the collated evidence, what is the current assessment of the effectiveness of interventions for natural system modification?	
Trade-offs between benefit and harms	• Use prescribed burning

Trade-off between benefit and harms

Use prescribed burning

Four studies in North America looked at bat activity and prescribed burning. One replicated, controlled, site comparison study found no difference in bat activity between burned and unburned forest. One replicated, site comparison study found higher activity of bat species that forage in the open in burned than unburned stands. One site comparison study found higher bat activity in forest preserves when prescribed burning was used with other restoration practices. One controlled, replicated, before-and-after study found that the home ranges of bats were closer to burned stands following fires. Four studies in North America (three replicated and one controlled) found bats roosting more often in burned areas, or equally in burned and unburned forest. *Assessment: trade-offs between benefits and harms (effectiveness 65%; certainty 50%; harms 20%).*

http://www.conservationevidence.com/actions/1006

2.9 Threat: Invasive species

2.9.1 Invasive species

Based on the collated evidence, what is the current assessment of the effectiveness of interventions for invasive species?	
Unknown effectiveness (limited evidence)	• Remove invasive plant species
Likely to be ineffective or harmful	• Translocate to predator or disease free areas
No evidence found (no assessment)	• Control invasive predators

Unknown effectiveness (limited evidence)

Remove invasive plant species

One site comparison study in North America found higher bat activity in forest preserves where invasive plant species had been removed alongside other restoration practices. *Assessment: unknown effectiveness (effectiveness 20%; certainty 10%; harms 0%).*

http://www.conservationevidence.com/actions/1008

Likely to be ineffective or harmful

● Translocate to predator or disease free areas

Two small unreplicated studies in New Zealand and Switzerland found low numbers of bats remaining at release sites after translocation, and observed homing tendencies, disease and death. *Assessment: Likely to be ineffective or harmful (effectiveness 5%; certainty 40%; harms 80%).*

http://www.conservationevidence.com/actions/1009

No evidence found (no assessment)

We have captured no evidence for the following interventions:

- Control invasive predators

2.9.2 White-nose syndrome

Based on the collated evidence, what is the current assessment of the effectiveness of interventions for white-nose syndrome?	
No evidence found (no assessment)	• Control anthropogenic spread • Cull infected bats • Increase population resistance • Modify cave environments to increase bat survival

No evidence found (no assessment)

We have captured no evidence for the following interventions:

- Control anthropogenic spread
- Cull infected bats
- Increase population resistance
- Modify cave environments to increase bat survival

2.10 Threat: Pollution

2.10.1 Domestic and urban waste water

Based on the collated evidence, what is the current assessment of the effectiveness of interventions for domestic and urban waste water?	
Unknown effectiveness (limited evidence)	• Change effluent treatments of domestic and urban waste water

Unknown effectiveness (limited evidence)

Change effluent treatments of domestic and urban waste water

We found no evidence for the effects on bats of changing effluent treatments of domestic and urban waste water discharged into rivers. One replicated, site comparison study in the UK found that foraging activity over filter bed sewage treatment works was higher than activity over active sludge systems. *Assessment: unknown effectiveness (effectiveness 40%; certainty 30%; harms 30%).*

http://www.conservationevidence.com/actions/1014

2.10.2 Agricultural and forestry effluents

Based on the collated evidence, what is the current assessment of the effectiveness of interventions for agricultural and forestry effluents?	
No evidence found (no assessment)	• Introduce legislation to control use of fertilizers, insecticides and pesticides • Change effluent treatments used in agriculture and forestry

No evidence found (no assessment)

We have captured no evidence for the following interventions:

- Introduce legislation to control use of fertilizers, insecticides and pesticides
- Change effluent treatments used in agriculture and forestry

2.10.3 Light and noise pollution

Based on the collated evidence, what is the current assessment of the effectiveness of interventions for light and noise pollution?	
Likely to be beneficial	• Leave bat roosts, roost entrances and commuting routes unlit • Minimize excess light pollution
No evidence found (no assessment)	• Restrict timing of lighting • Use low pressure sodium lamps or use UV filters • Impose noise limits in proximity to roosts and bat habitats

Likely to be beneficial

Leave bat roosts, roost entrances and commuting routes unlit

Two replicated studies in the UK found more bats emerging from roosts or flying along hedgerows when left unlit than when illuminated with white lights or streetlamps. *Assessment: likely to be beneficial (effectiveness 80%; certainty 50%; harms 0%).*

http://www.conservationevidence.com/actions/1017

Minimize excess light pollution

One replicated, randomized, controlled study in the UK found that bats avoided flying along hedgerows with dimmed lighting, and activity levels were lower than along unlit hedges. We found no evidence for the effects of reducing light spill using directional lighting or hoods on bats. *Assessment: likely to be beneficial (effectiveness 65%; certainty 50%; harms 0%).*

http://www.conservationevidence.com/actions/1018

No evidence found (no assessment)

We have captured no evidence for the following interventions:

- Restrict timing of lighting
- Use low pressure sodium lamps or use UV filters
- Impose noise limits in proximity to roosts and bat habitats

2.10.4 Timber treatments

Based on the collated evidence, what is the current assessment of the effectiveness of interventions for timber treatments?	
Beneficial	• Use mammal safe timber treatments in roof spaces
Likely to be ineffective or harmful	• Restrict timing of treatment

Beneficial

● Use mammal safe timber treatments in roof spaces

Two controlled laboratory studies in the UK found commercial timber treatments (containing lindane and pentachlorophenol) to be lethal to bats, but found alternative artificial insecticides (including permethrin) and three other fungicides did not increase bat mortality. Sealants over timber treatments had varying success. *Assessment: likely to be beneficial (effectiveness 90%; certainty 80%; harms 0%).*

http://www.conservationevidence.com/actions/1022

Likely to be ineffective or harmful

Restrict timing of treatment

One controlled laboratory experiment in the UK found that treating timber with lindane and pentachlorophenol 14 months prior to exposure by bats increased survival time but did not prevent death. Bats in cages treated with permethrin survived just as long when treatments were applied two months or 14 months prior to exposure. *Assessment: Likely to be ineffective or harmful (effectiveness 5%; certainty 55%; harms 50%).*

http://www.conservationevidence.com/actions/1023

2.11 Providing artificial roost structures for bats

Based on the collated evidence, what is the current assessment of the effectiveness of providing artificial roost structures for bats?	
Likely to be beneficial	• Provide artificial roost structures for bats

Likely to be beneficial

Provide artificial roost structures for bats

We found 22 replicated studies of artificial roost structures from across the world. Twenty-one studies show use of artificial roosts by bats. One study in the USA found that bats did not use the bat houses provided. Fifteen studies show varying occupancy rates of bats in artificial roost structures (3–100%). Two studies in Europe found an increase in bat populations using bat boxes in forest and woodland. Eight studies looked at bat box position. Three of four studies found that box orientation and exposure to sunlight are important for occupancy. Two studies found more bats occupying bat boxes on buildings than trees. Two studies found more bats occupying bat boxes in farm forestry or pine stands than in native or deciduous forest. Eleven studies looked at bat box design, including size, number of compartments and temperature, and found varying results. *Assessment: likely to be beneficial (effectiveness 60%; certainty 60%; harms 0%)*

http://www.conservationevidence.com/actions/1024

2.12 Education and awareness raising

Based on the collated evidence, what is the current assessment of the effectiveness of interventions for education and awareness raising?	
No evidence found (no assessment)	• Provide training to professionals • Educate homeowners about building and planning laws • Educate to improve public perception and raise awareness

No evidence found (no assessment)

We have captured no evidence for the following interventions:

- Provide training to professionals
- Educate homeowners about building and planning laws
- Educate to improve public perception and raise awareness

3. BIRD CONSERVATION

David R. Williams, Matthew F. Child, Lynn V. Dicks, Nancy Ockendon, Robert G. Pople, David A. Showler, Jessica C. Walsh, Erasmus K. H. J. zu Ermgassen & William J. Sutherland

Expert assessors

Tatsuya Amano, University of Cambridge, UK
Andy Brown, Natural England, UK
Fiona Burns, Royal Society for the Protection of Birds, UK
Yohay Carmel, Israel Institute of Technology
Mick Clout, University of Auckland, New Zealand
Geoff Hilton, Wildfowl & Wetlands Trust, UK
Nancy Ockendon, University of Cambridge, UK
James Pearce-Higgins, British Trust for Ornithology, UK
Sugoto Roy, Food and Environment Research Agency, DEFRA, UK
Rebecca K. Smith, University of Cambridge, UK
William J. Sutherland, University of Cambridge, UK
Judit Szabo, Charles Darwin University, Australia
Bernie Tershy, University of California, USA
Des Thomson, Scottish Natural Heritage, UK
Stuart Warrington, National Trust, UK
David Williams, University of Cambridge, UK

Scope of assessment: for native wild bird species across the world.

Assessed: 2015.

Effectiveness measure is the median % score.

Certainty measure is the median % certainty of evidence, determined by the quantity and quality of the evidence in the synopsis.

Harm measure is the median % score for negative side-effects to the group of species of concern. This was not scored for section 3.11 on invasive species.

 http://dx.doi.org/10.11647/OBP.0109.03

This book is meant as a guide to the evidence available for different conservation interventions and as a starting point in assessing their effectiveness. The assessments are based on the available evidence for the target group of species for each intervention. The assessment may therefore refer to different species or habitat to the one(s) you are considering. Before making any decisions about implementing interventions it is vital that you read the more detailed accounts of the evidence in order to assess their relevance for your study species or system.

Full details of the evidence are available at
www.conservationevidence.com

There may also be significant negative side-effects on the target groups or other species or communities that have not been identified in this assessment.

A lack of evidence means that we have been unable to assess whether or not an intervention is effective or has any harmful impacts.

3.1 Habitat protection

Based on the collated evidence, what is the current assessment of the effectiveness of interventions for habitat protection?	
Likely to be beneficial	• Legally protect habitats
Trade-offs between benefit and harms	• Provide or retain un-harvested buffer strips
Unknown effectiveness (limited evidence)	• Ensure connectivity between habitat patches

Likely to be beneficial

Legally protect habitats for birds

Four studies from Europe found that populations increased after habitat protection and a review from China found high use of protected habitats by cranes. A replicated, randomised and controlled study from Argentina found that some, but not all bird groups had higher species richness or were at higher densities in protected habitats. *Assessment: likely to be beneficial (effectiveness 50%; certainty 52%; harms 0%).*

http://www.conservationevidence.com/actions/158

Trade-off between benefit and harms

Provide or retain un-harvested buffer strips

Three replicated studies from the USA found that species richness or abundances were higher in narrow (<100 m) strips of forest, but five replicated studies from North America found that wider strips retained a community more similar to that of uncut forest than narrow strips. Tw replicated studies from the USA found no differences in productivity between wide and narrow buffers, but that predation of artificial nests was higher in buffers than in continuous forest. *Assessment: trade-offs between benefits and harms (effectiveness 60%; certainty 55%; harms 20%).*

http://www.conservationevidence.com/actions/161

Unknown effectiveness (limited evidence)

Ensure connectivity between habitat patches

Two studies of a replicated, controlled experiment in Canadian forests found that some species (not forest specialists) were found at higher densities in forest patches connected to continuous forest, compared to isolated patches and that some species used corridors more than clearcuts between patches. *Assessment: unknown effectiveness — limited evidence (effectiveness 38%; certainty 38%; harms 0%).*

http://www.conservationevidence.com/actions/160

3.2 Education and awareness raising

Based on the collated evidence, what is the current assessment of the effectiveness of interventions for education and awareness raising?	
Likely to be beneficial	• Raise awareness amongst the general public through campaigns and public information
Unknown effectiveness (limited evidence)	• Provide bird feeding materials to families with young children
No evidence found (no assessment)	• Enhance bird taxonomy skills through higher education and training • Provide training to conservationists and land managers on bird ecology and conservation

Likely to be beneficial

Raise awareness amongst the general public through campaigns and public information

A literature review from North America found that education was not sufficient to change behaviour, but that it was necessary for the success of economic incentives and law enforcement. *Assessment: likely to be beneficial (effectiveness 45%; certainty 48%; harms 0%).*

http://www.conservationevidence.com/actions/162

Unknown effectiveness (limited evidence)

Provide bird feeding materials to families with young children

A single replicated, paired study from the USA found that most children involved in a programme providing families with bird food increased their knowledge of birds, but did not significantly change their environmental attitudes. *Assessment: unknown effectiveness — limited evidence (effectiveness 42%; certainty 20%; harms 0%).*

http://www.conservationevidence.com/actions/163

No evidence found (no assessment)

We have captured no evidence for the following interventions:

- Enhance bird taxonomy skills through higher education and training
- Provide training to conservationists and land managers on bird ecology and conservation

3.3 Threat: Residential and commercial development

Based on the collated evidence, what is the current assessment of the effectiveness of interventions for residential and commercial development?	
Unknown effectiveness (limited evidence)	• Angle windows to reduce bird collisions • Mark windows to reduce bird collisions

Unknown effectiveness (limited evidence)

Angle windows to reduce bird collisions

A single randomised, replicated and controlled experiment in the USA found that fewer birds collided with windows angled away from the vertical. *Assessment: unknown effectiveness — limited evidence (effectiveness 60%; certainty 20%; harms 0%).*

http://www.conservationevidence.com/actions/166

Mark windows to reduce bird collisions

Two randomised, replicated and controlled studies found that marking windows did not appear to reduce bird collisions. However, when windows were largely covered with white cloth, or tinted, fewer birds flew towards or collided with them. A third randomised, replicated and controlled study found that fewer birds collided with tinted windows than with un-tinted ones, although the authors noted that the poor reflective quality of the glass could have influenced the results. *Assessment: unknown effectiveness — limited evidence (effectiveness 20%; certainty 20%; harms 0%).*

http://www.conservationevidence.com/actions/167

3.4 Threat: Agriculture

3.4.1 All farming systems

Based on the collated evidence, what is the current assessment of the effectiveness of interventions for all farming systems?	
Beneficial	• Plant wild bird seed or cover mixture • Provide (or retain) set-aside areas in farmland
Likely to be beneficial	• Create uncultivated margins around intensive arable or pasture fields • Increase the proportion of natural/semi-natural habitat in the farmed landscape • Manage ditches to benefit wildlife • Pay farmers to cover the costs of conservation measures • Plant grass buffer strips/margins around arable or pasture fields • Plant nectar flower mixture/wildflower strips • Leave refuges in fields during harvest • Reduce conflict by deterring birds from taking crops: use bird scarers • Relocate nests at harvest time to reduce nestling mortality • Use mowing techniques to reduce mortality
Unknown effectiveness (limited evidence)	• Control scrub on farmland • Offer per clutch payment for farmland birds • Manage hedges to benefit wildlife • Plant new hedges • Reduce conflict by deterring birds from taking crops: use repellents • Take field corners out of management

Likely to be ineffective or harmful	• Mark bird nests during harvest or mowing
No evidence found (no assessment)	• Cross compliance standards for all subsidy payments • Food labelling schemes relating to biodiversity-friendly farming • Manage stone-faced hedge banks to benefit birds • Plant in-field trees • Protect in-field trees • Reduce field size (or maintain small fields) • Support or maintain low-intensity agricultural systems • Tree pollarding, tree surgery

Beneficial

Plant wild bird seed or cover mixture

Seven of 41 studies found that fields or farms with wild bird cover had higher diversity than other sites, or that wild bird cover held more species than other habitats. Thirty-two studies found that populations, or abundances of some or all species were higher on wild bird cover than other habitats, or that wild bird cover was used more than other habitats. Four of these studies investigated several interventions at once. Thirteen studies found that bird populations or densities were similar on wild bird cover and other habitats that some species were not associated with wild bird cover, or that birds rarely used wild bird cover. Three studies found higher productivities of birds on wild bird cover than other habitats. Two found no differences for some or all species studied. Two studies found that survival of grey partridge or artificial nests increased on wild bird cover; one found lower partridge survival in farms with wild bird cover than other farms. Five studies from the UK found that some wild bird cover crops were used more than others. A study and a review found that the arrangement of wild bird cover in the landscape affected its use by birds. *Assessment: beneficial (effectiveness 81%; certainty 81%; harms 0%).*

http://www.conservationevidence.com/actions/187

Provide (or retain) set-aside areas in farmland

Four out of 23 studies from Europe and North America found more species on set-aside than on crops. One study found fewer. Twenty-one studies found that some species were at higher densities on set-aside than other habitats, or that they used set-aside more often. Four found that some species were found at lower densities on set-aside than other habitats. Three studies found that waders and Eurasian skylarks had higher productivities on set-aside than other crops. One study found that skylarks on set-aside had lower similar or lower productivities than on crops. One study from the UK found that rotational set-aside was used more than non-rotational set-aside, another found no difference. A review from North America and Europe found that naturally regenerated set-aside held more birds and more species than sown set-aside. *Assessment: beneficial (effectiveness 70%; certainty 75%; harms 0%).*

http://www.conservationevidence.com/actions/175

Likely to be beneficial

Create uncultivated margins around intensive arable or pasture fields

One of eight studies found that three sparrow species found on uncultivated margins on a site in the USA were not found on mown field edges. A replicated study from Canada found fewer species in uncultivated margins than in hedges or trees. Three studies found that some bird species were associated with uncultivated margins, or that birds were more abundant on margins than other habitats. One study found that these effects were very weak and four studies of three experiments found that uncultivated margins contained similar numbers of birds as other habitats in winter, or that several species studied did not show associations with margins. A study from the UK found that yellowhammers used uncultivated margins more than crops in early summer. Use fell in uncut margins later in the year. A study from the UK found that grey partridge released on uncultivated margins had high survival. *Assessment: likely to be beneficial (effectiveness 45%; certainty 55%; harms 0%).*

http://www.conservationevidence.com/actions/190

Increase the proportion of natural/semi-natural habitat in the farmed landscape

Two studies from Switzerland and Australia, of the five we captured, found that areas with plantings of native species, or areas under a scheme designed to increase semi-natural habitats (the Swiss Ecological Compensation Areas scheme), held more bird species than other areas. One study from Switzerland found that populations of three bird species increased in areas under the Ecological Compensation Areas scheme. A third Swiss study found that some habitats near Ecological Compensation Areas held more birds than habitats further away, but the overall amount of Ecological Compensation Area had no effect on bird populations. A study from the UK found no effect of habitat-creation on grey partridge populations. *Assessment: likely to be beneficial (effectiveness 45%; certainty 44%; harms 0%).*

http://www.conservationevidence.com/actions/171

Manage ditches to benefit wildlife

One study of four from the UK found that bunded ditches were visited more often by birds than non-bunded ditches. Three studies found that some birds responded positively to ditches managed for wildlife, but that other species did not respond to management, or responded negatively. *Assessment: likely to be beneficial (effectiveness 40%; certainty 49%; harms 14%).*

http://www.conservationevidence.com/actions/180

Pay farmers to cover the costs of conservation measures

Three out of 31 studies found national population increases in three species after payment schemes targeted at their conservation. One found that many other species continued declining. Twenty-two studies found that at least some species were found at higher densities on sites with agri-environment schemes; some differences were present only in summer or only in winter. Fifteen studies found some species at similar densities on agri-environment schemes and non-agri-environment scheme sites or appeared to respond negatively to agri-environment schemes. One study found that grey partridge survival was higher in some years on agri-environment scheme sites. Two studies found higher productivity on agri-environment scheme sites for some species, one found no effect of agri-environment schemes. A review found that some agri-environment schemes options were not being used enough to benefit many species of bird. A study from the UK found that there was no difference in the densities of seed-eating birds in winter between two agri-environment scheme designations. *Assessment: likely to be beneficial (effectiveness 56%; certainty 80%; harms 0%).*

http://www.conservationevidence.com/actions/172

Plant grass buffer strips/margins around arable or pasture fields

One of 15 studies found more bird species in fields in the USA that were bordered by grass margins than in unbordered fields. Two studies from the UK found no effect of margins on species richness. One study found that more birds used grass strips in fields than used crops. Even more used grass margins. Nine studies from the USA and UK found that sites with grass margins had more positive population trends or higher populations for some birds, or that some species showed strong habitat associations with grass margins. Three studies found no such effect for some or all species. Two studies found that species used margins more than other habitats and one found that birds used cut margins more than uncut during winter, but less than other habitats during summer. A study from the UK found that grey partridge broods were smaller on grass margins than other habitat types. *Assessment: likely to be beneficial (effectiveness 47%; certainty 54%; harms 0%).*

http://www.conservationevidence.com/actions/191

Plant nectar flower mixture/wildflower strips

Three of seven studies found that birds used wildflower strips more than other habitats; two found strips were not used more than other habitats. A study from Switzerland found that Eurasian skylarks were more likely to nest in patches sown with annual weeds than in crops and were less likely to abandon nests. A study from the UK found that management of field margins affected their use more than the seed mix used. *Assessment: likely to be beneficial (effectiveness 55%; certainty 45%; harms 0%).*

http://www.conservationevidence.com/actions/189

Leave refuges in fields during harvest

One study found that fewer gamebirds came into contact with mowing machinery when refuges were left in fields. A study from the UK found that Eurasian skylarks did not nest at higher densities in uncut refuges than in the rest of the field. *Assessment: likely to be beneficial (effectiveness 50%; certainty 41%; harms 0%).*

http://www.conservationevidence.com/actions/193

Reduce conflict by deterring birds from taking crops (using bird scarers)

A controlled paired study in the USA found reduced levels of damage to almond orchards when American crow distress calls were broadcast. A study in Pakistan found that four pest species were less abundant when reflector ribbons were hung above crops compared to where ribbons were not used. *Assessment: likely to be beneficial (effectiveness 66%; certainty 44%; harms 0%).*

http://www.conservationevidence.com/actions/199

Relocate nests at harvest time to reduce nestling mortality

A study from Spain found that Montagu's harrier clutches had higher hatching and fledging rates when they were temporarily moved during harvest than control nests that were not moved. *Assessment: likely to be beneficial (effectiveness 55%; certainty 42%; harms 0%).*

http://www.conservationevidence.com/actions/195

Use mowing techniques to reduce mortality

One of three studies from the UK found a large increase in the national population of corncrakes after a scheme to delay mowing and promote corncrake-friendly mowing techniques. Two studies found lower levels of corncrake and Eurasian skylark mortality when wildlife-friendly mowing techniques were used. *Assessment: likely to be beneficial (effectiveness 85%; certainty 50%; harms 0%).*

http://www.conservationevidence.com/actions/192

Unknown effectiveness (limited evidence)

Control scrub on farmland

A study from the UK found farms with a combined intervention that included scrub control had lower numbers of young grey partridge per adult. *Assessment: unknown effectiveness — limited evidence (effectiveness 7%; certainty 9%; harms 1%).*

http://www.conservationevidence.com/actions/197

Offer per clutch payment for farmland birds

One of two studies from the Netherlands found slightly higher breeding densities of waders on farms with per clutch payment schemes but this and another study found no higher numbers overall. One study found higher hatching success on farms with payment schemes. *Assessment: unknown effectiveness — limited evidence (effectiveness 43%; certainty 35%; harms 0%).*

http://www.conservationevidence.com/actions/196

Manage hedges to benefit wildlife

One of seven studies found no differences in the number of species in a UK site with wildlife-friendly hedge management and sites without. Seven studies found that some species increased in managed hedges or were more likely to be found in them than other habitats. One investigated several interventions at the same time. Four studies found that some species responded negatively or not at all to hedge management or that effects varied across regions of the UK. *Assessment: unknown effectiveness — limited evidence (effectiveness 39%; certainty 38%; harms 3%).*

http://www.conservationevidence.com/actions/177

Plant new hedges

A study from the USA found that populations of northern bobwhites increased following several interventions including the planting of new hedges. *Assessment: unknown effectiveness — limited evidence (effectiveness 23%; certainty 19%; harms 0%).*

http://www.conservationevidence.com/actions/178

Reduce conflict by deterring birds from taking crops (using repellents)

A replicated, randomised and controlled *ex situ* study in the USA found that dickcissels consumed less rice if it was treated with two repellents compared to controls. *Assessment: unknown effectiveness — limited evidence (effectiveness 29%; certainty 27%; harms 0%).*

http://www.conservationevidence.com/actions/200

Take field corners out of management

A study from the UK found that overwinter survival of grey partridge was positively correlated with taking field corners out of management, but this relationship was only significant in one of three winters. There was no relationship with measures of productivity (brood size, young: adult). *Assessment: unknown effectiveness — limited evidence (effectiveness 30%; certainty 15%; harms 0%).*

http://www.conservationevidence.com/actions/198

Likely to be ineffective or harmful

Mark bird nests during harvest or mowing

A study from the Netherlands found that fewer northern lapwing nests were destroyed when they were marked with bamboo poles than when they were unmarked. *Assessment: likely to be ineffective or harmful (effectiveness 30%; certainty 45%; harms 20%).*

http://www.conservationevidence.com/actions/148

No evidence found (no assessment)

We have captured no evidence for the following interventions:

- Cross compliance standards for all subsidy payments
- Food labelling schemes relating to biodiversity-friendly farming
- Manage stone-faced hedge banks to benefit birds
- Plant in-field trees
- Protect in-field trees
- Reduce field size (or maintain small fields)
- Support or maintain low-intensity agricultural systems
- Tree pollarding, tree surgery

3.4.2 Arable farming

Based on the collated evidence, what is the current assessment of the effectiveness of interventions for arable farming systems?	
Likely to be beneficial	• Create 'skylark plots' • Leave overwinter stubbles • Leave uncropped cultivated margins or fallow land (includes lapwing and stone curlew plots) • Sow crops in spring rather than autumn • Undersow spring cereals, with clover for example
Trade-off between benefit and harms	• Reduce tillage
Unknown effectiveness (limited evidence)	• Implement mosaic management • Increase crop diversity to benefit birds • Plant more than one crop per field (intercropping)
Unlikely to be beneficial	• Create beetle banks
Likely to be ineffective or harmful	• Plant cereals in wide-spaced rows • Revert arable land to permanent grassland
No evidence found (no assessment)	• Add 1% barley into wheat crop for corn buntings • Create corn bunting plots • Leave unharvested cereal headlands within arable fields • Plant nettle strips

Likely to be beneficial

Create 'skylark plots' (undrilled patches in cereal fields)

One study of seven found that the Eurasian skylark population on a farm increased after skylark plots were provided. Another found higher skylark densities on fields with plots in. Two studies from the UK found that skylark productivity was higher for birds with skylark plots in their territories, a study from Switzerland found no differences. Two studies from Denmark and Switzerland found that skylarks used plots more than expected, but a study from the UK found that seed-eating songbirds did not. *Assessment: likely to be beneficial (effectiveness 65%; certainty 60%; harms 0%).*

http://www.conservationevidence.com/actions/214

Leave overwinter stubbles

Three of fourteen studies report positive population-level changes in two species after winter stubble provision. All investigated several interventions at once. Eight studies found that some farmland birds were found on stubbles or were positively associated with them, three investigated several interventions and one found no more positive associations than expected by chance. A study from the UK found that most species did not preferentially use stubble, compared to cover crops and another found that a greater area of stubble in a site meant lower grey partridge brood size. Five studies from the UK found that management of stubbles influenced their use by birds. One study found that only one species was more common on stubbles under agri-environment schemes. *Assessment: likely to be beneficial (effectiveness 40%; certainty 60%; harms 0%).*

http://www.conservationevidence.com/actions/203

Leave uncropped cultivated margins or fallow land (includes lapwing and stone curlew plots)

Three of nine studies report that the UK population of Eurasian thick-knees increased following a scheme to promote lapwing plots (and other interventions). A study from the UK found that plots did not appear to influence grey partridge populations. Four studies from the UK found that at least one species was associated with lapwing plots, or used them for foraging or nesting. One study found that 11 species were not associated with plots, another that fewer used plots than used crops in two regions of the UK. Two studies found that nesting success was higher on lapwing plots and fallow than in crops. A third found fewer grey partridge chicks per adult on sites with lots of lapwing plots. *Assessment: likely to be beneficial (effectiveness 59%; certainty 55%; harms 15%).*

http://www.conservationevidence.com/actions/213

Sow crops in spring rather than autumn

One study from Sweden, of three examining the effects of spring-sown crops, found that more birds were found on areas with spring, rather than autumn-sown crops. A study from the UK found that several species used the study site for the first time after spring-sowing was started. All three studies found that some populations increased after the start of spring sowing. A study from the UK found that some species declined as well. A study from Sweden found that hatching success of songbirds and northern lapwing was lower on spring-sown, compared with autumn-sown crops. *Assessment: likely to be beneficial (effectiveness 55%; certainty 67%; harms 10%).*

http://www.conservationevidence.com/actions/207

Undersow spring cereals, with clover for example

Four of five studies from the UK found that bird densities were higher on undersown fields or margins than other fields, or that use of fields increased if they were undersown. Two studies of the same experiment found that not all species nested at higher densities in undersown habitats. A study from the UK found that grey partridge populations were lower on sites with large amounts of undersown cereal. *Assessment: likely to be beneficial (effectiveness 60%; certainty 45%; harms 10%).*

http://www.conservationevidence.com/actions/208

Trade-off between benefit and harms

Reduce tillage

Six of ten studies found that some or all bird groups had higher species richness or diversity on reduced-tillage fields, compared to conventional fields in some areas. Two studies found that some groups had lower diversity on reduced-tillage sites, or that there was no difference between treatments. Nine studies found that some species were found at higher densities on reduced tillage fields, six found that some species were at similar or lower densities. Three studies found evidence for higher productivities on reduced-tillage fields. One found that not all measures of productivity were higher. *Assessment: trade-offs between benefits and harms (effectiveness 50%; certainty 48%; harms 51%).*

http://www.conservationevidence.com/actions/211

Unknown effectiveness (limited evidence)

Implement mosaic management

One of two studies from the Netherlands found that northern lapwing population trends, but not those of three other waders, became more positive following the introduction of mosaic management. The other found that black-tailed godwit productivity was higher under mosaic management than other management types. *Assessment: unknown effectiveness — limited evidence (effectiveness 20%; certainty 33%; harms 0%).*

http://www.conservationevidence.com/actions/130

Increase crop diversity to benefit birds

A study from the UK found that more barnacle geese used a site after the amount of land under cereals was decreased and several other interventions were used. *Assessment: unknown effectiveness — limited evidence (effectiveness 20%; certainty 19%; harms 0%).*

http://www.conservationevidence.com/actions/201

Plant more than one crop per field (intercropping)

A study from the USA found that 35 species of bird used fields with intercropping, with four nesting, but that productivity from the fields was very low. *Assessment: unknown effectiveness — limited evidence (effectiveness 30%; certainty 36%; harms 18%).*

http://www.conservationevidence.com/actions/209

Unlikely to be beneficial

Create beetle banks

Two of six studies from the UK found that some bird populations were higher on sites with beetle banks. Both investigated several interventions at once. Two studies found no relationships between bird species abundances or populations and beetle banks. Two studies (including a review) from the

UK found that three bird species used beetle banks more than expected, one used them less than expected. *Assessment: unlikely to be beneficial (effectiveness 30%; certainty 41%; harms 0%).*

http://www.conservationevidence.com/actions/217

Likely to be ineffective or harmful

Plant cereals in wide-spaced rows

One of three studies from the UK found that fields with wide-spaced rows held more Eurasian skylark nests than control fields. One study found that fields with wide-spaced rows held fewer nests. Both found that fields with wide-spaced rows held fewer nests than fields with skylark plots. A study from the UK found that skylark chicks in fields with wide-spaced rows had similar diets to those in control fields. *Assessment: likely to be ineffective or harmful (effectiveness 20%; certainty 44%; harms 20%).*

http://www.conservationevidence.com/actions/216

Revert arable land to permanent grassland

All five studies looking at the effects of reverting arable land to grassland found no clear benefit to birds. The studies monitored birds in winter or grey partridges in the UK and wading birds in Denmark. They included three replicated controlled trials. *Assessment: likely to be ineffective or harmful (effectiveness 0%; certainty 64%; harms 10%).*

http://www.conservationevidence.com/actions/210

No evidence found (no assessment)

We have captured no evidence for the following interventions:

- Add 1% barley into wheat crop for corn buntings
- Create corn bunting plots
- Leave unharvested cereal headlands within arable fields
- Plant nettle strips

3.4.3 Livestock farming

Based on the collated evidence, what is the current assessment of the effectiveness of interventions for livestock farming systems?	
Likely to be beneficial	• Delay mowing date on grasslands • Leave uncut rye grass in silage fields • Maintain species-rich, semi-natural grassland • Maintain traditional water meadows • Mark fencing to avoid bird mortality • Plant cereals for whole crop silage • Reduce grazing intensity • Reduce management intensity of permanent grasslands
Trade-off between benefit and harms	• Exclude livestock from semi-natural habitat
Unknown effectiveness (limited evidence)	• Create open patches or strips in permanent grassland • Maintain upland heath/moor • Protect nests from livestock to reduce trampling • Provide short grass for waders • Raise mowing height on grasslands
Unlikely to be beneficial	• Use traditional breeds of livestock
No evidence found (no assessment)	• Maintain lowland heathland • Maintain rush pastures • Maintain wood pasture and parkland • Plant Brassica fodder crops • Use mixed stocking

Likely to be beneficial

Delay mowing date on grasslands

Two of five studies (both reviews) found that the UK corncrake populations increased following two schemes to encourage farmers to delay mowing. A study from the Netherlands found no evidence that waders and other birds were more abundant in fields with delayed mowing. Another study from the Netherlands found that fields with delayed mowing held more birds than other fields, but differences were present before the scheme began and population trends did not differ between treatments. A study from the USA found that fewer nests were destroyed by machinery in late-cut fields, compared with early-cut fields. *Assessment: likely to be beneficial (effectiveness 45%; certainty 52%; harms 0%).*

http://www.conservationevidence.com/actions/223

Leave uncut rye grass in silage fields

All four studies from the UK (including two reviews) found that seed-eating birds were benefited by leaving uncut (or once-cut) rye grass in fields, or that seed-eating species were more abundant on uncut plots. Three studies found that seed-eating birds were more abundant on uncut and ungrazed plots than on uncut and grazed plots. A study from the UK found that the responses of non-seed-eating birds were less certain than seed-eating species, with some species avoiding uncut rye grass. *Assessment: likely to be beneficial (effectiveness 67%; certainty 56%; harms 8%).*

http://www.conservationevidence.com/actions/224

Maintain species-rich, semi-natural grassland

One of two studies found that the populations of five species increased in an area of the UK after the start of management designed to maintain unimproved grasslands. A study from Switzerland found that wetland birds nested at greater densities on managed hay meadows than expected, but birds of open farmland used hay meadows less. *Assessment: likely to be beneficial (effectiveness 41%; certainty 44%; harms 0%).*

http://www.conservationevidence.com/actions/218

Maintain traditional water meadows

One of four studies (from the UK) found that the populations of two waders increased on reserves managed as water meadows. Two studies from the Netherlands found that there were more waders or birds overall on specially managed meadows or 12.5 ha plots, but one found that these differences were present before management began, the other found no differences between individual fields under different management. Two studies from the UK and Netherlands found that wader populations were no different between specially and conventionally managed meadows, or that wader populations decreased on specially-managed meadows. A study from the UK found that northern lapwing productivity was not high enough to maintain populations on three of four sites managed for waders. *Assessment: likely to be beneficial (effectiveness 50%; certainty 52%; harms 0%).*

http://www.conservationevidence.com/actions/229

Mark fencing to avoid bird mortality

A study from the UK found that fewer birds collided with marked sections of deer fences, compared to unmarked sections. *Assessment: likely to be beneficial (effectiveness 65%; certainty 46%; harms 0%).*

http://www.conservationevidence.com/actions/238

Plant cereals for whole crop silage

Three studies of one experiment found that seed-eating birds used cereal-based wholecrop silage crops more than other crops in summer and winter. Insect-eating species used other crops and grassland more often. *Assessment: likely to be beneficial (effectiveness 55%; certainty 43%; harms 0%).*

http://www.conservationevidence.com/actions/225

Reduce grazing intensity

Nine of eleven studies from the UK and USA found that the populations of some species were higher on fields with reduced grazing intensity, compared to conventionally-grazed fields, or found that birds used these fields more. Three studies investigated several interventions at once. Five studies from Europe found that some or all species were no more numerous, or were less abundant on fields with reduced grazing. A study from the UK found that black grouse populations increased at reduced grazing sites (whilst they declined elsewhere). However, large areas with reduced grazing had low female densities. A study from the USA found that the number of species on plots with reduced grazing increased over time. A study from four European countries found no differences in the number of species on sites with low- or high-intensity grazing. *Assessment: likely to be beneficial (effectiveness 46%; certainty 55%; harms 0%).*

http://www.conservationevidence.com/actions/220

Reduce management intensity of permanent grasslands

Seven of eight European studies found that some or all birds studied were more abundant on grasslands with reduced management intensity, or used them more than other habitats for foraging. Five studies of four experiments found that some or all species were found at lower or similar abundances on reduced-management grasslands, compared to intensively-managed grasslands. *Assessment: likely to be beneficial (effectiveness 65%; certainty 46%; harms 0%).*

http://www.conservationevidence.com/actions/219

Trade-off between benefit and harms

Exclude livestock from semi-natural habitat

Two studies from the USA, out of 11 overall, found higher species richness on sites with grazers excluded. A study from Argentina found lower species richness and one from the USA found no difference. Seven studies from the USA found that overall bird abundance, or the abundances of some species were higher in sites with grazers excluded. Seven studies from the USA and Argentina found that overall abundance or the abundance of some species were lower on sites without grazers, or did not differ. Three studies found that productivities were higher on sites with grazers excluded. In one, the difference was only found consistently in comparison with improved pastures, not unimproved. *Assessment: trade-offs between benefits and harms (effectiveness 50%; certainty 57%; harms 30%).*

http://www.conservationevidence.com/actions/236

Unknown effectiveness (limited evidence)

Create open patches or strips in permanent grassland

A study from the UK found that Eurasian skylarks used fields with open strips in, but that variations in skylark numbers were too great to draw conclusions from this finding. *Assessment: unknown effectiveness — limited evidence (effectiveness 20%; certainty 19%; harms 0%).*

http://www.conservationevidence.com/actions/239

Maintain upland heath/moor

A study from the UK found that bird populations in one region were increasing with agri-environment guidelines on moor management. There were some problems with overgrazing, burning and scrub encroachment. *Assessment: unknown effectiveness — limited evidence (effectiveness 30%; certainty 15%; harms 0%).*

http://www.conservationevidence.com/actions/230

Protect nests from livestock to reduce trampling

One of two studies found that a population of Chatham Island oystercatchers increased following several interventions including the erection of fencing around individual nests. A study from Sweden found that no southern dunlin nests were trampled when protected by cages; some unprotected nests were destroyed. *Assessment: unknown effectiveness — limited evidence (effectiveness 56%; certainty 19%; harms 0%).*

http://www.conservationevidence.com/actions/237

Provide short grass for waders

A study from the UK found that common starlings and northern lapwing spent more time foraging on areas with short swards, compared to longer swards. *Assessment: unknown effectiveness — limited evidence (effectiveness 41%; certainty 32%; harms 0%).*

http://www.conservationevidence.com/actions/221

Raise mowing height on grasslands

One of two studies from the UK found that no more foraging birds were attracted to plots with raised mowing heights, compared to plots with shorter grass. A review from the UK found that Eurasian skylarks had higher productivity on sites with raised mowing heights, but this increase was not enough to maintain local populations. *Assessment: unknown effectiveness — limited evidence (effectiveness 20%; certainty 36%; harms 0%).*

http://www.conservationevidence.com/actions/222

Unlikely to be beneficial

Use traditional breeds of livestock

A study from four countries in Europe found no differences in bird abundances in areas grazed with traditional or commercial breeds. *Assessment: unlikely to be beneficial (effectiveness 0%; certainty 44%; harms 0%).*

http://www.conservationevidence.com/actions/233

No evidence found (no assessment)

We have captured no evidence for the following interventions:

- Maintain lowland heathland
- Maintain rush pastures
- Maintain wood pasture and parkland
- Plant Brassica fodder crops
- Use mixed stocking

3.4.4 Perennial, non-timber crops

Based on the collated evidence, what is the current assessment of the effectiveness of interventions for perennial, non-timber crops?	
Unknown effectiveness (limited evidence)	• Maintain traditional orchards
No evidence found (no assessment)	• Manage perennial bioenergy crops to benefit wildlife

Unknown effectiveness (limited evidence)

Maintain traditional orchards

Two site comparison studies from the UK and Switzerland found that traditional orchards offer little benefit to birds. In Switzerland only one breeding bird species was associated with traditional orchards. In the UK, the population density of cirl bunting was negatively related to the presence of orchards. *Assessment: unknown effectiveness — limited evidence (effectiveness 10%; certainty 24%; harms 0%).*

http://www.conservationevidence.com/actions/240

No evidence found (no assessment)

We have captured no evidence for the following interventions:

- Manage perennial bioenergy crops to benefit wildlife

3.4.5 Aquaculture

Based on the collated evidence, what is the current assessment of the effectiveness of interventions for aquaculture?	
Likely to be beneficial	• Deter birds from landing on shellfish culture gear suspend oyster bags under water • Deter birds from landing on shellfish culture gear use spikes on oyster cages • Disturb birds at roosts • Provide refuges for fish within ponds • Use electric fencing to exclude fish-eating birds • Use 'mussel socks' to prevent birds from attacking shellfish • Use netting to exclude fish-eating birds
Unknown effectiveness (limited evidence)	• Increase water turbidity to reduce fish predation by birds • Translocate birds away from fish farms • Use in-water devices to reduce fish loss from ponds
Unlikely to be beneficial	• Disturb birds using foot patrols • Spray water to deter birds from ponds
Likely to be ineffective or harmful	• Scare birds from fish farms

Likely to be beneficial

Deter birds from landing on shellfish culture gear

A study from Canada found that fewer birds landed on oyster cages fitted with spikes than control cages. The same study found that fewer birds landed on oyster bags suspended 6 cm, but not 3 cm, underwater, compared to bags on the surface. *Assessment for using spikes on oyster cages: likely to be beneficial (effectiveness 60%; certainty 43%; harms 0%). Assessment for suspending oyster bags under water: likely to be beneficial (effectiveness 55%; certainty 43%; harms 0%).*

http://www.conservationevidence.com/actions/257
http://www.conservationevidence.com/actions/256

Disturb birds at roosts

One study from the USA found reduced fish predation after fish-eating birds were disturbed at roosts. Five studies from the USA and Israel found that birds foraged less near disturbed roosts, or left the area after being disturbed. One found the effects were only temporary. *Assessment: likely to be beneficial (effectiveness 67%; certainty 45%; harms 0%).*

http://www.conservationevidence.com/actions/245

Provide refuges for fish within ponds

A study from the UK found that cormorants caught fewer fish in a pond with fish refuges in, compared to a control pond. *Assessment: likely to be beneficial (effectiveness 65%; certainty 43%; harms 0%).*

http://www.conservationevidence.com/actions/253

Use electric fencing to exclude fish-eating birds

Two before-and-after trials from the USA found lower use of fish ponds by herons after electric fencing was installed. *Assessment: likely to be beneficial (effectiveness 60%; certainty 49%; harms 0%).*

http://www.conservationevidence.com/actions/247

Use 'mussel socks' to prevent birds from attacking shellfish

A study from Canada found that mussel socks with protective sleeves lost fewer medium-sized mussels (but not small or large mussels), compared to unprotected mussel socks. *Assessment: likely to be beneficial (effectiveness 50%; certainty 41%; harms 0%).*

http://www.conservationevidence.com/actions/250

Use netting to exclude fish-eating birds

Two studies from Germany and the USA, and a review, found that netting over ponds reduced the loss of fish to predatory birds. Two studies from the USA and the Netherlands found that birds still landed on ponds with netting, but that they altered their behaviour, compared to open ponds. Two studies from Germany and Israel found that some birds became entangled in netting over ponds. *Assessment: likely to be beneficial (effectiveness 60%; certainty 59%; harms 15%).*

http://www.conservationevidence.com/actions/248

Unknown effectiveness (limited evidence)

Increase water turbidity to reduce fish predation by birds

An *ex situ* study from France found that egret foraging efficiency was reduced in more turbid water. *Assessment: unknown effectiveness — limited evidence (effectiveness 50%; certainty 23%; harms 0%).*

http://www.conservationevidence.com/actions/252

Translocate birds away from fish farms

A study from the USA found that translocating birds appeared to reduce bird numbers at a fish farm. A study from Belgium found that it did not. *Assessment: unknown effectiveness — limited evidence (effectiveness 20%; certainty 33%; harms 0%).*

http://www.conservationevidence.com/actions/251

Use in-water devices to reduce fish loss from ponds

A study from the USA found that fewer cormorants used two ponds after underwater ropes were installed; a study from Australia found that no fewer cormorants used ponds with gill nets in. *Assessment: unknown effectiveness — limited evidence (effectiveness 34%; certainty 35%; harms 0%).*

http://www.conservationevidence.com/actions/254

Unlikely to be beneficial

Disturb birds using foot patrols

Two replicated studies from Belgium and Australia found that using foot patrols to disturb birds from fish farms did not reduce the number of birds present or fish consumption. *Assessment: unlikely to be beneficial (effectiveness 0%; certainty 45%; harms 0%).*

http://www.conservationevidence.com/actions/249

Spray water to deter birds from ponds

A study from Sweden found that spraying water deterred birds from fish ponds, but that some birds became habituated to the spray. *Assessment: unlikely to be beneficial (effectiveness 31%; certainty 43%; harms 0%).*

http://www.conservationevidence.com/actions/255

Likely to be ineffective or harmful

Scare birds from fish farms

One study from Israel found a population increase in fish-eating birds after efforts to scare them from fish farms, possibly due to lower persecution. One of two studies found evidence for reduced loss of fish when birds were scared from farms. Two studies from Australia and Belgium found that disturbing birds using foot patrols was not effective. Ten of 11 studies from across the world found some effects for acoustic deterrents, five of seven found that visual deterrents were effective. In both cases some studies found that results were temporary, birds became habituated or that some deterrents were effective, whilst others were not. One study found that trained raptors were effective, one found little evidence for the effectiveness of helicopters or light aircraft. *Assessment: likely to be ineffective or harmful (effectiveness 36%; certainty 64%; harms 0%).*

http://www.conservationevidence.com/actions/244

3.5 Threat: Energy production and mining

Based on the collated evidence, what is the current assessment of the effectiveness of interventions for energy production and mining?	
Unknown effectiveness (limited evidence)	• Paint wind turbines to increase their visibility

Unknown effectiveness (limited evidence)

Paint wind turbines to increase their visibility

A single, controlled *ex situ* experiment found that thick black stripes running across a wind turbine's blades made them more conspicuous to an American kestrel *Falco sparverius* than control (unpatterned) blades. Other designs were less visible or indistinguishable from controls. *Assessment: unknown effectiveness — limited evidence (effectiveness 16%; certainty 10%; harms 0%).*

http://www.conservationevidence.com/actions/258

3.6 Threat: Transportation and service corridors

3.6.1 Verges and airports

Based on the collated evidence, what is the current assessment of the effectiveness of interventions for verges and airports?	
Likely to be beneficial	• Scare or otherwise deter birds from airports
Unknown effectiveness (limited evidence)	• Mow roadside verges
No evidence found (no assessment)	• Sow roadside verges

Likely to be beneficial

Scare or otherwise deter birds from airports

Two replicated studies in the UK and USA found that fewer birds used areas of long grass at airports, but no data were provided on the effect of long grass on strike rates or bird mortality. *Assessment: likely to be beneficial (effectiveness 50%; certainty 44%; harms 0%).*

http://www.conservationevidence.com/actions/261

Unknown effectiveness (limited evidence)

Mow roadside verges

A single replicated, controlled trial in the USA found that mowed roadside verges were less attractive to ducks as nesting sites, but had higher nesting success after four years. *Assessment: unknown effectiveness — limited evidence (effectiveness 30%; certainty 30%; harms 9%).*

http://www.conservationevidence.com/actions/259

No evidence found (no assessment)

We have captured no evidence for the following intervention:

- Sow roadside verges

3.6.2 Power lines and electricity pylons

Based on the collated evidence, what is the current assessment of the effectiveness of interventions for power lines and electricity pylons?	
Beneficial	• Mark power lines
Likely to be beneficial	• Bury or isolate power lines • Insulate electricity pylons • Remove earth wires from power lines • Use perch-deterrents to stop raptors perching on pylons
Unknown effectiveness (limited evidence)	• Thicken earth wires
Unlikely to be beneficial	• Add perches to electricity pylons • Reduce electrocutions by using plastic, not metal, leg rings to mark birds • Use raptor models to deter birds from power lines

Beneficial

Mark power lines

A total of eight studies and two literature reviews from across the world found that marking power lines led to significant reductions in bird collision mortalities. Different markers had different impacts. *Assessment: beneficial (effectiveness 81%; certainty 85%; harms 0%).*

http://www.conservationevidence.com/actions/265

Likely to be beneficial

Bury or isolate power lines

A single before-and-after study in Spain found a dramatic increase in juvenile eagle survival following the burial or isolation of dangerous power lines. *Assessment: likely to be beneficial (effectiveness 60%; certainty 44%; harms 0%).*

http://www.conservationevidence.com/actions/262

Insulate electricity pylons

A single before-and-after study in the USA found that insulating power pylons significantly reduced the number of Harris's hawks electrocuted. *Assessment: likely to be beneficial (effectiveness 60%; certainty 45%; harms 0%).*

http://www.conservationevidence.com/actions/268

Remove earth wires from power lines

Two before-and-after studies from Norway and the USA describe significant reductions in bird collision mortalities after earth wires were removed from sections of power lines. *Assessment: likely to be beneficial (effectiveness 90%; certainty 60%; harms 0%).*

http://www.conservationevidence.com/actions/263

Use perch-deterrents to stop raptors perching on pylons

A single controlled study in the USA found that significantly fewer raptors were found near perch-deterrent lines, compared to controls, but no information on electrocutions was provided. *Assessment: likely to be beneficial (effectiveness 50%; certainty 45%; harms 0%).*

http://www.conservationevidence.com/actions/269

Unknown effectiveness (limited evidence)

Thicken earth wires

A single paired sites trial in the USA found no reduction in crane species collision rates in a wire span with an earth wire three times thicker than normal. *Assessment: unknown effectiveness — limited evidence (effectiveness 0%; certainty 25%; harms 0%).*

http://www.conservationevidence.com/actions/264

Unlikely to be beneficial

Add perches to electricity pylons

A single before-and-after study in Spain found that adding perches to electricity pylons did not reduce electrocutions of Spanish imperial eagles. *Assessment: unlikely to be beneficial (effectiveness 0%; certainty 42%; harms 0%).*

http://www.conservationevidence.com/actions/267

Reduce electrocutions by using plastic, not metal, leg rings to mark birds

A single replicated and controlled study in the USA found no evidence that using plastic leg rings resulted in fewer raptors being electrocuted. *Assessment: unlikely to be beneficial (effectiveness 0%; certainty 42%; harms 0%).*

http://www.conservationevidence.com/actions/270

Use raptor models to deter birds from power lines

A single paired sites trial in Spain found that installing raptor models near power lines had no impact on bird collision mortalities. *Assessment: unlikely to be beneficial (effectiveness 0%; certainty 43%; harms 0%)*

http://www.conservationevidence.com/actions/266

3.7 Threat: Biological resource use

3.7.1 Reducing exploitation and conflict

Based on the collated evidence, what is the current assessment of the effectiveness of interventions for reducing exploitation and conflict?	
Beneficial	• Use legislative regulation to protect wild populations
Likely to be beneficial	• Use wildlife refuges to reduce hunting disturbance
Unknown effectiveness (limited evidence)	• Employ local people as 'biomonitors' • Increase 'on-the-ground' protection to reduce unsustainable levels of exploitation • Introduce voluntary 'maximum shoot distances' • Mark eggs to reduce their appeal to collectors • Move fish-eating birds to reduce conflict with fishermen • Promote sustainable alternative livelihoods • Provide 'sacrificial grasslands' to reduce conflict with farmers • Relocate nestlings to reduce poaching • Use education programmes and local engagement to help reduce persecution or exploitation of species
No evidence found (no assessment)	• Use alerts during shoots to reduce mortality of non-target species

Scare fish-eating birds from areas to reduce conflict

Studies investigating scaring fish from fishing areas are discussed in 'Threat: Agriculture — Aquaculture'.

Beneficial

Use legislative regulation to protect wild populations

Five out of six studies from Europe, Asia, North America and across the world, found evidence that stricter legislative protection was correlated with increased survival, lower harvests or increased populations. The sixth, a before-and-after study from Australia, found that legislative protection did not reduce harvest rates. *Assessment: beneficial (effectiveness 65%; certainty 65%; harms 0%).*

http://www.conservationevidence.com/actions/271

Likely to be beneficial

Use wildlife refuges to reduce hunting disturbance

Three studies from the USA and Europe found that more birds used refuges where hunting was not allowed, compared to areas with hunting, and more used the refuges during the open season. However, no studies examined the population-level effects of refuges. *Assessment: likely to be beneficial (effectiveness 45%; certainty 45%; harms 0%).*

http://www.conservationevidence.com/actions/278

Unknown effectiveness (limited evidence)

Employ local people as 'biomonitors'

A single replicated study in Venezuela found that poaching of parrot nestlings was significantly lower in years following the employment of five local people as 'biomonitors'. *Assessment: unknown effectiveness — limited evidence (effectiveness 50%; certainty 19%; harms 0%).*

http://www.conservationevidence.com/actions/275

Increase 'on-the-ground' protection to reduce unsustainable levels of exploitation

Two before-and-after studies from Europe and Central America found increases in bird populations and recruitment following stricter anti-poaching methods or the stationing of a warden on the island in question. However, the increases in Central America were only short-term, and were lost when the intensive effort was reduced. *Assessment: unknown effectiveness — limited evidence (effectiveness 50%; certainty 25%; harms 0%).*

http://www.conservationevidence.com/actions/272

Introduce voluntary 'maximum shoot distances'

A single study from Denmark found a significant reduction in the injury rates of pink-footed geese following the implementation of a voluntary maximum shooting distance. *Assessment: unknown effectiveness — limited evidence (effectiveness 40%; certainty 20%; harms 0%).*

http://www.conservationevidence.com/actions/279

Mark eggs to reduce their appeal to collectors

A single before-and-after study in Australia found increased fledging success of raptor eggs in a year they were marked with a permanent pen. *Assessment: unknown effectiveness — limited evidence (effectiveness 50%; certainty 35%; harms 0%).*

http://www.conservationevidence.com/actions/276

Move fish-eating birds to reduce conflict with fishermen

A single before-and-after study in the USA found that Caspian tern chicks had a lower proportion of commercial fish in their diet following the movement of the colony away from an important fishery. *Assessment: unknown effectiveness — limited evidence (effectiveness 32%; certainty 24%; harms 0%).*

http://www.conservationevidence.com/actions/281

Promote sustainable alternative livelihoods

A single before-and-after study in Costa Rica found that a scarlet macaw population increased following several interventions including the promotion of sustainable, macaw-based livelihoods. *Assessment: unknown effectiveness — limited evidence (effectiveness 30%; certainty 19%; harms 0%).*

http://www.conservationevidence.com/actions/273

Provide 'sacrificial grasslands' to reduce conflict with farmers

Two UK studies found that more geese used areas of grassland managed for them, but that this did not appear to attract geese from outside the study site and therefore was unlikely to reduce conflict with farmers. *Assessment: unknown effectiveness — limited evidence (effectiveness 18%; certainty 20%; harms 0%).*

http://www.conservationevidence.com/actions/280

Relocate nestlings to reduce poaching

A single replicated study in Venezuela found a significant reduction in poaching rates and an increase in fledging rates of yellow-shouldered amazons when nestlings were moved into police premises overnight. *Assessment: unknown effectiveness — limited evidence (effectiveness 50%; certainty 30%; harms 0%).*

http://www.conservationevidence.com/actions/277

Use education programmes and local engagement to help reduce persecution or exploitation of species

Six out of seven studies from across the world found increases in bird populations or decreases in mortality following education programmes, whilst one study from Venezuela found no evidence that poaching decreased following an educational programme. In all but one study reporting successes, other interventions were also used, and a literature review from the USA and Canada argues that education was not sufficient to change behaviour, although a Canadian study found that there was a significant shift in local peoples' attitudes to conservation and exploited species following educational programmes. *Assessment: unknown effectiveness — limited evidence (effectiveness 50%; certainty 30%; harms 0%).*

http://www.conservationevidence.com/actions/274

No evidence found (no assessment)

We have captured no evidence for the following interventions:

- Use alerts during shoots to reduce mortality of non-target species

3.7.2 Reducing fisheries bycatch

Based on the collated evidence, what is the current assessment of the effectiveness of interventions for reducing fisheries bycatch?	
Beneficial	• Use streamer lines to reduce seabird bycatch on longlines
Likely to be beneficial	• Mark trawler warp cables to reduce seabird collisions • Reduce seabird bycatch by releasing offal overboard when setting longlines • Weight baits or lines to reduce longline bycatch of seabirds
Trade-off between benefit and harms	• Set lines underwater to reduce seabird bycatch • Set longlines at night to reduce seabird bycatch
Unknown effectiveness (limited evidence)	• Dye baits to reduce seabird bycatch • Thaw bait before setting lines to reduce seabird bycatch • Turn deck lights off during night-time setting of longlines to reduce bycatch • Use a sonic scarer when setting longlines to reduce seabird bycatch • Use acoustic alerts on gillnets to reduce seabird bycatch • Use bait throwers to reduce seabird bycatch • Use bird exclusion devices such as 'Brickle curtains' to reduce seabird mortality when hauling longlines • Use high visibility mesh on gillnets to reduce seabird bycatch • Use shark liver oil to deter birds when setting lines
Likely to be ineffective or harmful	• Use a line shooter to reduce seabird bycatch
No evidence found (no assessment)	• Reduce bycatch through seasonal or area closures • Reduce 'ghost fishing' by lost/discarded gear • Reduce gillnet deployment time to reduce seabird bycatch • Set longlines at the side of the boat to reduce seabird bycatch

	• Tow buoys behind longlining boats to reduce seabird bycatch • Use a water cannon when setting longlines to reduce seabird bycatch • Use high-visibility longlines to reduce seabird bycatch • Use larger hooks to reduce seabird bycatch on longlines

Beneficial

Use streamer lines to reduce seabird bycatch on longlines

Ten studies from coastal and pelagic fisheries across the globe found strong evidence for reductions in bycatch when streamer lines were used. Five studies from the South Atlantic, New Zealand and Australia were inconclusive, uncontrolled or had weak evidence for reductions. One study from the sub-Antarctic Indian Ocean found no evidence for reductions. Three studies from around the world found that bycatch rates were lower when two streamers were used compared to one, and one study found rates were lower still with three streamers. *Assessment: beneficial (effectiveness 65%; certainty 75%; harms 0%).*

http://www.conservationevidence.com/actions/285

Likely to be beneficial

Mark trawler warp cables to reduce seabird collisions

A single replicated and controlled study in Argentina found lower seabird mortality (from colliding with warp cables) when warp cables were marked with orange traffic cones. *Assessment: likely to be beneficial (effectiveness 54%; certainty 40%; harms 0%).*

http://www.conservationevidence.com/actions/305

Reduce seabird bycatch by releasing offal overboard when setting longlines

Two replicated and controlled studies in the South Atlantic and sub-Antarctic Indian Ocean found significantly lower seabird bycatch rates when offal was released overboard as lines were being set. *Assessment: likely to be beneficial (effectiveness 51%; certainty 50%; harms 0%).*

http://www.conservationevidence.com/actions/299

Weight baits or lines to reduce longline bycatch of seabirds

Three replicated and controlled studies from the Pacific found lower bycatch rates of some seabird species on weighted longlines. An uncontrolled study found low bycatch rates with weighted lines but that weights only increased sink rates in small sections of the line. Some species were found to attack weighted lines more than control lines. *Assessment: likely to be beneficial (effectiveness 46%; certainty 45%; harms 15%).*

http://www.conservationevidence.com/actions/296

Trade-off between benefit and harms

Set lines underwater to reduce seabird bycatch

Five studies in Norway, South Africa and the North Pacific found lower seabird bycatch rates on longlines set underwater. However, results were species-specific, with shearwaters and possibly albatrosses continuing to take baits set underwater. *Assessment: trade-offs between benefits and harms (effectiveness 61%; certainty 50%; harms 24%).*

http://www.conservationevidence.com/actions/288

Set longlines at night to reduce seabird bycatch

Six out of eight studies from around the world found lower bycatch rates when longlines were set at night, but the remaining two found higher bycatch rates (of northern fulmar in the North Pacific and white-chinned petrels in the South Atlantic, respectively). Knowing whether bycatch species are night- or day-feeding is therefore important in reducing bycatch rates. *Assessment: trade-offs between benefits and harms (effectiveness 60%; certainty 70%; harms 48%).*

http://www.conservationevidence.com/actions/283

Unknown effectiveness (limited evidence)

Dye baits to reduce seabird bycatch

A single randomised, replicated and controlled trial in Hawaii, USA, found that albatrosses attacked baits at significantly lower rates when baits were dyed blue. *Assessment: unknown effectiveness — limited evidence (effectiveness 50%; certainty 20%; harms 0%).*

http://www.conservationevidence.com/actions/293

Thaw bait before setting lines to reduce seabird bycatch

A study from Australia found that longlines set using thawed baits caught significantly fewer seabirds than controls. *Assessment: unknown effectiveness — limited evidence (effectiveness 50%; certainty 30%; harms 0%).*

http://www.conservationevidence.com/actions/298

Turn deck lights off during night-time setting of longlines to reduce bycatch

A single replicated and controlled study in the South Atlantic found lower seabird bycatch rates on night-set longlines when deck lights were turned off. *Assessment: unknown effectiveness — limited evidence (effectiveness 51%; certainty 21%; harms 0%).*

http://www.conservationevidence.com/actions/284

Use a sonic scarer when setting longlines to reduce seabird bycatch

A single study from the South Atlantic found that seabirds only temporarily changed behaviour when a sonic scarer was used, and seabird bycatch rates did not appear to be lower on lines set with a scarer. *Assessment: unknown effectiveness — limited evidence (effectiveness 2%; certainty 10%; harms 0%).*

http://www.conservationevidence.com/actions/295

Use acoustic alerts on gillnets to reduce seabird bycatch

A randomised, replicated and controlled trial in a coastal fishery in the USA found that fewer guillemots (common murres) but not rhinoceros auklets were caught in gillnets fitted with sonic alerts. *Assessment: unknown effectiveness — limited evidence (effectiveness 44%; certainty 21%; harms 0%).*

http://www.conservationevidence.com/actions/301

Use bait throwers to reduce seabird bycatch

A single analysis found significantly lower seabird bycatch on Australian longliners when a bait thrower was used to set lines. *Assessment: unknown effectiveness — limited evidence (effectiveness 46%; certainty 30%; harms 0%).*

http://www.conservationevidence.com/actions/291

Use bird exclusion devices such as 'Brickle curtains' to reduce seabird mortality when hauling longlines

A single replicated study found that Brickle curtains reduced the number of seabirds caught, when compared to an exclusion device using only a single boom. Using purse seine buoys as well as the curtain appeared to be even more effective, but sample sizes did not allow useful comparisons. *Assessment: unknown effectiveness — limited evidence (effectiveness 48%; certainty 30%; harms 0%).*

http://www.conservationevidence.com/actions/302

Use high visibility mesh on gillnets to reduce seabird bycatch

A single randomised, replicated and controlled trial in a coastal fishery in the USA found that fewer guillemots (common murres) and rhinoceros auklets were caught in gillnets with higher percentages of brightly coloured netting. However, such netting also reduced the catch of the target salmon. *Assessment: unknown effectiveness — limited evidence (effectiveness 60%; certainty 30%; harms 0%).*

http://www.conservationevidence.com/actions/303

Use shark liver oil to deter birds when setting lines

Two out of three replicated and controlled trials in New Zealand found that fewer birds followed boats or dived for baits when non-commercial shark oil was dripped off the back of the boat. *Assessment: unknown effectiveness — limited evidence (effectiveness 30%; certainty 25%; harms 0%).*

http://www.conservationevidence.com/actions/297

Likely to be ineffective or harmful

Use a line shooter to reduce seabird bycatch

Two randomised, replicated and controlled trials found that seabird bycatch rates were higher (in the North Pacific) or the same (in Norway) on longlines set using line shooters, compared to those set without a shooter. *Assessment: likely to be ineffective or harmful (effectiveness 0%; certainty 50%; harms 40%).*

http://www.conservationevidence.com/actions/290

No evidence found (no assessment)

We have captured no evidence for the following interventions:

- Reduce bycatch through seasonal or area closures
- Reduce 'ghost fishing' by lost/discarded gear
- Reduce gillnet deployment time to reduce seabird bycatch
- Set longlines at the side of the boat to reduce seabird bycatch
- Tow buoys behind longlining boats to reduce seabird bycatch
- Use a water cannon when setting longlines to reduce seabird bycatch
- Use high-visibility longlines to reduce seabird bycatch
- Use larger hooks to reduce seabird bycatch on longlines

3.8 Threat: Human intrusions and disturbance

Based on the collated evidence, what is the current assessment of the effectiveness of interventions for human intrusions and disturbance?	
Likely to be beneficial	• Provide paths to limit disturbance • Start educational programmes for personal watercraft owners • Use signs and access restrictions to reduce disturbance at nest sites • Use voluntary agreements with local people to reduce disturbance
Unknown effectiveness (limited evidence)	• Habituate birds to human visitors • Use nest covers to reduce the impact of research on predation of ground-nesting seabirds
No evidence found (no assessment)	• Reduce visitor group sizes • Set minimum distances for approaching birds (buffer zones)

Likely to be beneficial

Provide paths to limit disturbance

A study from the UK found that two waders nested closer to a path, or at higher densities near the path, following resurfacing, which resulted in far fewer people leaving the path. *Assessment: likely to be beneficial (effectiveness 50%; certainty 40%; harms 0%).*

http://www.conservationevidence.com/actions/311

Start educational programmes for personal watercraft owners

A before-and-after study in the USA found that common tern reproduction increased, and rates of disturbance decreased, following a series of educational programmes aimed at recreational boat users. *Assessment: likely to be beneficial (effectiveness 40%; certainty 40%; harms 0%).*

http://www.conservationevidence.com/actions/314

Use signs and access restrictions to reduce disturbance at nest sites

Six studies from across the world found increased numbers of breeders, higher reproductive success or lower levels of disturbance in waders and terns following the start of access restrictions or the erection of signs near nesting areas. Two studies from Europe and Antarctica found no effect of access restrictions on reproductive success in eagles and penguins, respectively. *Assessment: likely to be beneficial (effectiveness 59%; certainty 55%; harms 10%).*

http://www.conservationevidence.com/actions/309

Use voluntary agreements with local people to reduce disturbance

A before-and-after trial in the USA found significantly lower rates of waterfowl disturbance following the establishment of a voluntary waterfowl avoidance area, despite an overall increase in boat traffic. *Assessment: likely to be beneficial (effectiveness 50%; certainty 40%; harms 0%).*

http://www.conservationevidence.com/actions/313

Unknown effectiveness (limited evidence)

Habituate birds to human visitors

A study from Australia found that bridled terns from heavily disturbed sites had similar or higher reproductive success compared with less-disturbed sites, possibly suggesting that habituation had occurred. *Assessment: unknown effectiveness — limited evidence (effectiveness 20%; certainty 10%; harms 0%).*

http://www.conservationevidence.com/actions/315

Use nest covers to reduce the impact of research on predation of ground-nesting seabirds

A before-and-after study from Canada found that hatching success of Caspian terns was significantly higher when researchers protected nests after disturbing adults from them. *Assessment: unknown effectiveness — limited evidence (effectiveness 41%; certainty 35%; harms 19%).*

http://www.conservationevidence.com/actions/316

No evidence found (no assessment)

We have captured no evidence for the following interventions:

- Reduce visitor group sizes
- Set minimum distances for approaching birds (buffer zones)

3.9 Threat: Natural system modifications

Based on the collated evidence, what is the current assessment of the effectiveness of interventions for natural system modifications?	
Likely to be beneficial	• Create scrapes and pools in wetlands and wet grasslands • Provide deadwood/snags in forests: use ring-barking, cutting or silvicides • Use patch retention harvesting instead of clearcutting
Trade-off between benefit and harms	• Clear or open patches in forests • Employ grazing in artificial grassland/pastures • Employ grazing in natural grasslands • Employ grazing in non-grassland habitats • Manage water level in wetlands • Manually control or remove midstorey and ground-level vegetation: forests • Manually control or remove midstorey and ground-level vegetation: mow or cut natural grasslands • Manually control or remove midstorey and ground-level vegetation: mow or cut semi-natural grasslands/pastures • Manually control or remove midstorey and ground-level vegetation: shrubland • Raise water levels in ditches or grassland • Thin trees within forests

	• Use prescribed burning: grasslands • Use prescribed burning: pine forests • Use prescribed burning: savannahs • Use prescribed burning: shrublands • Use selective harvesting/logging instead of clearcutting
Unknown effectiveness (limited evidence)	• Clearcut and re-seed forests • Coppice trees • Fertilise grasslands • Manage woodland edges for birds • Manually control or remove midstorey and ground-level vegetation: reedbeds • Manually control or remove midstorey and ground-level vegetation: savannahs • Plant trees to act as windbreaks • Plough habitats • Provide deadwood/snags in forests: add woody debris to forests • Remove coarse woody debris from forests • Replace non-native species of tree/shrub • Re-seed grasslands • Use environmentally sensitive flood management • Use fire suppression/control • Use greentree reservoir management • Use prescribed burning: Australian sclerophyll forest • Use shelterwood cutting instead of clearcutting • Use variable retention management during forestry operations
Likely to be ineffective or harmful	• Apply herbicide to mid- and understorey vegetation • Treat wetlands with herbicides • Use prescribed burning: coastal habitats • Use prescribed burning: deciduous forests
No evidence found (no assessment)	• Protect nest trees before burning

Likely to be beneficial

Create scrapes and pools in wetlands and wet grasslands

Four out of six studies from the UK and North America found that more bird used sites, or breeding populations on sites increased, after ponds or scrapes were created. A study from the USA found that some duck species used newly created ponds and others used older ponds. A study from the UK found that northern lapwing chicks foraged in newly created features and that chick condition was higher in sites with a large number of footdrains. *Assessment: likely to be beneficial (effectiveness 75%; certainty 60%; harms 0%).*

http://www.conservationevidence.com/actions/359

Provide deadwood/snags in forests (use ring-barking, cutting or silvicides)

One of five studies found that forest plots provided with snags had higher bird diversity and abundance than plots without snags. Three of four studies from the USA and UK found that species used artificially-created snags for nesting and foraging. One study from the USA found that use increased with how long a snag had been dead. *Assessment: likely to be beneficial (effectiveness 45%; certainty 50%; harms 0%).*

http://www.conservationevidence.com/actions/343

Use patch retention harvesting instead of clearcutting

One of two studies (from the USA) found that areas under patch retention harvesting contained more birds of more species than clearcut areas, retaining similar numbers to unharvested areas. Two studies found that forest specialist species were found more frequently in patch retention plots than under other management. Habitat generalists declined on patch retention sites compared to other managements. *Assessment: likely to be beneficial (effectiveness 70%; certainty 46%; harms 0%).*

http://www.conservationevidence.com/actions/330

Trade-off between benefit and harms

Clear or open patches in forests

Seven out of nine studies from the UK and USA found that early-successional species increased in clearcut areas of forests, compared to other management. Two studies found that mature-forest species declined. One study found no differences in species richness between treatments, another found no consistent differences. A study from the USA found that a mosaic of cut and uncut areas supported a variety of species. *Assessment: trade-offs between benefits and harms (effectiveness 55%; certainty 60%; harms 30%).*

http://www.conservationevidence.com/actions/326

Employ grazing in artificial grasslands/pastures

Five studies from the UK and USA found use or nesting densities were higher in grazed compared to ungrazed areas. A study from Canada found an increase in duck populations following the start of grazing along with other interventions. Eight studies from the UK, Canada and the USA found species richness, community composition, abundances, use, nesting densities, nesting success or productivity were similar or lower on grazed compared with ungrazed areas. One found that several species were excluded by grazing. *Assessment: trade-offs between benefits and harms (effectiveness 43%; certainty 65%; harms 45%).*

http://www.conservationevidence.com/actions/349

Employ grazing in natural grasslands

Five of 12 studies from the USA and Canada found that densities of some species were higher on grazed than ungrazed sites. Eight studies from the USA, Canada and France found that some or all species studied were found at similar or lower densities on grazed compared to ungrazed sites or those under other management. Two controlled studies from the USA and Canada found that nesting success was higher on grazed than ungrazed sites. Five studies from the USA and Canada found that nesting success was similar or lower on grazed sites. *Assessment: trade-offs between benefits and harms (effectiveness 40%; certainty 60%; harms 50%).*

http://www.conservationevidence.com/actions/348

Employ grazing in non-grassland habitats

One of eight studies found more bird species on grazed than unmanaged sites, apart from in drought years. A study from the Netherlands found the number of species in a mixed habitat wetland site declined with increased grazing. Three studies in Sweden, the Netherlands and Kenya found that the overall abundance or densities of some species were higher in grazed than ungrazed sites. Four studies in Europe and Kenya found that some species were absent or at lower densities on grazed compared to ungrazed sites or those under different management. Five studies from across the world found no differences in abundances or densities of some or all species between grazed sites and those that were ungrazed or under different management. Two studies from the UK found that productivity was lower in grazed than ungrazed sites. A study from the UK found that songbirds and invertebrate-eating species, but not crows were more common on rough-grazed habitats than intensive pasture. *Assessment: trade-offs between benefits and harms (effectiveness 40%; certainty 67%; harms 40%).*

http://www.conservationevidence.com/actions/350

Manage water level in wetlands

Three studies (of six) from the USA, UK and Canada found that different species were more abundant at different water heights. One found that diversity levels also changed. One study found that great bitterns in the UK established territories earlier when deep water levels were maintained, but productivity did not vary. A study from Spain found that water management successfully retained water near a greater flamingo nesting area, but did not measure the effects on productivity or survival. *Assessment: trade-offs between benefits and harms (effectiveness 40%; certainty 41%; harms 35%).*

http://www.conservationevidence.com/actions/355

Manually control or remove midstorey and ground-level vegetation (including mowing, chaining, cutting etc.) (forests)

Seven studies from Europe and the USA found that species richness, total density or densities of some species were higher in areas with mid- or understorey management compared to areas without management. Four studies also used other interventions. Seven studies from the USA and

Canada found that species richness, densities, survival or competition for nest sites were similar or lower in areas with mid- or understorey control. Two studies investigated several interventions at once. Two studies from Canada found higher nest survival in forests with removal of deciduous trees compared to controls. One study found that chicks foraging success was higher in areas with cleared understorey vegetation compared to burned areas, but lower than under other managements. *Assessment: trade-offs between benefits and harms (effectiveness 40%; certainty 75%; harms 40%).*

http://www.conservationevidence.com/actions/335

Manually control or remove midstorey and ground-level vegetation (including mowing, chaining, cutting etc.) (mowing or cutting natural grasslands)

Two of six studies found higher densities of birds or nests on mown grasslands compared to unmanaged or burned areas. Two studies found lower densities or nests of some species and two found no differences in nesting densities or community composition on mown compared to unmown areas. One study from the USA found that grasshopper sparrow nesting success was higher on mown than grazed areas. One study from the USA found that duck nesting success was similar on cut and uncut areas. *Assessment: trade-offs between benefits and harms (effectiveness 40%; certainty 50%; harms 39%).*

http://www.conservationevidence.com/actions/338

Manually control or remove midstorey and ground-level vegetation (including mowing, chaining, cutting etc.) (mowing or cutting semi-natural grasslands/pastures)

One of four studies found that wader populations increased following annual cutting of semi-natural grasslands. One study from the UK found that ducks grazed at higher densities on cut areas. Another study in the UK found that goose grazing densities were unaffected by cutting frequency. One study from the USA found that Henslow's sparrows were more likely to be recaptured on unmown than mown grasslands. *Assessment: trade-offs between benefits and harms (effectiveness 40%; certainty 40%; harms 20%).*

http://www.conservationevidence.com/actions/339

Manually control or remove midstorey and ground-level vegetation (including mowing, chaining, cutting etc.) (shrublands)

One of seven studies found that overall bird diversity and bird density was similar between chained areas, burned areas and controls. One found that overall diversity and abundance was lower on mown sites than controls, but that grassland-specialist species were present on managed sites. Five studies from the USA and Europe found than some species were at greater densities or abundances on sites with mechanical vegetation control than on sites with burning or no management. Three studies from the USA found that some species were less abundant on sites with mechanical vegetation removal. One study from the USA found no differences between areas cut in winter and summer. *Assessment: trade-offs between benefits and harms (effectiveness 43%; certainty 54%; harms 30%).*

http://www.conservationevidence.com/actions/337

Raise water levels in ditches or grassland

One of seven studies found that three waders were found to have recolonised a UK site or be found at very high densities after water levels were raised. Three studies from Europe found that raising water levels on grassland provided habitat for waders. A study from Denmark found that oystercatchers did not nest at higher densities on sites with raised water levels. A study from the UK found that birds visited sites with raised water levels more frequently than other fields, but another UK study found that feeding rates did not differ between sites with raised water levels and those without. A study from the USA found that predation rates on seaside sparrow nests increased as water levels were raised. *Assessment: trade-offs between benefits and harms (effectiveness 65%; certainty 55%; harms 25%).*

http://www.conservationevidence.com/actions/354

Thin trees within forests

One study of 14 (from the USA) found higher bird species richness in sites with tree thinning and several other interventions, compared to unmanaged sites. Three studies from the UK and USA found no such differences. Seven studies (four investigating multiple interventions) found that overall bird abundance or the abundance of some species was higher in thinned plots, compared to those under different management. Five studies found that found that abundances were similar, or that some species were less abundant in areas with thinning. Two studies from the USA found no effect of thinning on wood thrushes, a species thought to be sensitive to it. A study from the USA found that a higher proportion of nests were in nest boxes in a thinned site, compared to a control. A study from the USA found no differences in bird abundances between burned sites with high-retention thinning, compared to low-retention sites. *Assessment: trade-offs between benefits and harms (effectiveness 50%; certainty 60%; harms 30%).*

http://www.conservationevidence.com/actions/328

Use prescribed burning (grasslands)

Four of 21 studies found that overall species richness and community composition did not vary between burned and unburned sites. Nine studies from across the world found that at least some species were more abundant or at higher densities in burned than unburned areas or areas under different management. Fourteen studies found that at least one species was at similar or lower abundances on burned areas. Responses varied depending on how soon after fires monitoring occurred. One study from the USA found that Florida grasshopper sparrow had significantly higher reproductive success soon after burns, whilst another found that dickcissel reproductive success was higher in patch-burned than burned and grazed areas. *Assessment: trade-offs between benefits and harms (effectiveness 45%; certainty 60%; harms 40%).*

http://www.conservationevidence.com/actions/322

Use prescribed burning (pine forests)

Four of 28 studies in the USA found higher species richness, densities or abundances in sites with prescribed burning, tree thinning and in one case mid- or understorey control compared to controls. Fourteen studies found that some species were more abundant, or had higher productivities or survival in burned or burned and thinned areas than control areas. One study found that effects varied with geography and habitat. Fifteen studies found no differences in species richness or densities, community composition, productivity, behaviour or survival between sites with prescribed burning or burning and thinning, and controls or sites with other management. One study found that foraging success of chicks was lower in burned areas. Three studies found effects did not vary with burn season. *Assessment: trade-offs between benefits and harms (effectiveness 50%; certainty 77%; harms 35%).*

http://www.conservationevidence.com/actions/318

Use prescribed burning (savannahs)

One of five studies found that burned areas of savannah tended to have more birds and species than control or grazed areas, although burned sites showed significant annual variation unlike grazed sites. A study from Australia found that effects on bird abundances depended on burn season and habitat type. Two studies in the USA found that some open country species were more common in burned areas than unburned. A study from the USA found that two eastern bluebirds successfully raised chicks after a local prescribed burn. *Assessment: trade-offs between benefits and harms (effectiveness 40%; certainty 50%; harms 35%).*

http://www.conservationevidence.com/actions/320

Use prescribed burning (shrublands)

One of eight studies found that overall bird densities were similar between burned and unburned areas, whilst another found that species numbers and densities did not vary between areas burned in summer or winter. Three studies found that some species were more abundant on areas that were burned. Four found that species densities were similar or lower on burned compared to control areas or those under different management. One study found that sage sparrows chose different nest sites before and after burning. Another found no differences in greater sage grouse movement between burned and unburned areas. *Assessment: trade-offs between benefits and harms (effectiveness 43%; certainty 50%; harms 45%).*

http://www.conservationevidence.com/actions/321

Use selective harvesting/logging instead of clearcutting

Six of seven studies from the USA and Canada found that some species were more, and other less, abundant in selectively logged forests compared to unlogged stands, or those under other management. One study found that differences between treatments were not consistent. A study from the USA found that species richness of cavity-nesting birds was lower in selectively logged forests than in clearcuts. One study from the USA found that brood parasitism was higher in selectively logged forests for two species and lower for two others, compared to control stands. *Assessment: trade-offs between benefits and harms (effectiveness 65%; certainty 60%; harms 30%).*

http://www.conservationevidence.com/actions/331

Unknown effectiveness (limited evidence)

Clearcut and re-seed forests

One of two studies from the USA found that stands of pines replanted with native species held more species typical of scrub habitats than stands under different management. The other study found similar bird densities in clearcut and re-seeded sites and those under different management. *Assessment: unknown effectiveness — limited evidence (effectiveness 30%; certainty 35%; harms 0%).*

http://www.conservationevidence.com/actions/327

Coppice trees

One of three studies found a population increase in European nightjars on a UK site after the introduction of coppicing and other interventions. Two studies from the UK and USA found that the use of coppices by some bird species declined over time. A UK study found that species richness decreased with the age of a coppice, but that some species were more abundant in older stands. *Assessment: unknown effectiveness — limited evidence (effectiveness 34%; certainty 30%; harms 30%).*

http://www.conservationevidence.com/actions/329

Fertilise grasslands

All four studies captured (all from the UK) found that more geese grazed on fertilised areas of grass more than control areas. Two investigated cutting and fertilizing at the same time. One study found that fertilised areas were used less than re-seeded areas. One study found that fertilisation had an effect at applications of 50 kg N/ha, but not at 18 kg N/ha. Another found that the effects of fertilisation did not increase at applications over 80 kg N/

ha. *Assessment: unknown effectiveness — limited evidence (effectiveness 60%; certainty 35%; harms 7%).*

http://www.conservationevidence.com/actions/353

Manage woodland edges for birds

One of three studies found that a local population of European nightjars increased at a UK site following the start of a management regime that included the management of woodland edges for birds. Two studies of an experiment in the USA found that bird abundance (but not species richness or nesting success) was higher in woodland edges managed for wildlife than unmanaged edges. *Assessment: unknown effectiveness — limited evidence (effectiveness 55%; certainty 39%; harms 30%).*

http://www.conservationevidence.com/actions/334

Manually control or remove midstorey and ground-level vegetation (including mowing, chaining, cutting etc.) (reedbeds)

One of three studies found that warblers nested at lower densities in cut areas of reeds. Productivity and success did not vary between treatments. A study from Denmark found that geese grazed at the highest densities on reedbeds cut 5–12 years previously. One study in the UK found that cutting reeds and changing water levels did not affect great bittern breeding productivity, but did delay territory establishment. *Assessment: unknown effectiveness — limited evidence (effectiveness 15%; certainty 36%; harms 14%).*

http://www.conservationevidence.com/actions/340

Manually control or remove midstorey and ground-level vegetation (including mowing, chaining, cutting etc.) (savannahs)

A study in Argentina found that in summer, but not overall, bird abundance and species richness was lower in an area where shrubs were removed compared to a control. Community composition also differed between treatments. *Assessment: unknown effectiveness — limited evidence (effectiveness 30%; certainty 10%; harms 30%).*

http://www.conservationevidence.com/actions/336

Plant trees to act as windbreaks

One of two studies found that a population of European nightjars increased at a UK site after multiple interventions including the planting of windbreak trees. A study from the USA found that such trees appeared to disrupt lekking behaviour in greater prairie chickens. *Assessment: unknown effectiveness — limited evidence (effectiveness 12%; certainty 25%; harms 20%).*

http://www.conservationevidence.com/actions/351

Plough habitats

One of four studies found that bird densities were higher on ploughed wetlands in the USA than unploughed ones. Three studies of one experiment in the UK found that few whimbrels nested on areas of heathland ploughed and re-seeded, but that they were used for foraging in early spring. There were no differences in chick survival between birds that used ploughed and re-seeded heathland and those that did not. *Assessment: unknown effectiveness — limited evidence (effectiveness 25%; certainty 36%; harms 10%).*

http://www.conservationevidence.com/actions/358

Provide deadwood/snags in forests (adding woody debris to forests)

One study from Australia found that brown treecreeper numbers were higher in plots with large amounts of dead wood added compared to plots with less or no debris added. *Assessment: unknown effectiveness — limited evidence (effectiveness 50%; certainty 29%; harms 0%).*

http://www.conservationevidence.com/actions/344

Remove coarse woody debris from forests

Two studies from the USA found that some species increased in sites with woody debris removal. One found that overall breeding bird abundance and diversity were lower in removal plots; the other that survival of black-chinned hummingbird nests was lower. *Assessment: unknown effectiveness — limited evidence (effectiveness 10%; certainty 33%; harms 60%).*

http://www.conservationevidence.com/actions/345

Replace non-native species of tree/shrub

A study from the USA found that the number of black-chinned hummingbird nests increased after fuel reduction and the planting of native species, but that the increase was smaller than at sites without planting. *Assessment: unknown effectiveness — limited evidence (effectiveness 5%; certainty 18%; harms 0%).*

http://www.conservationevidence.com/actions/341

Re-seed grasslands

One of two studies from the UK found that geese grazed at higher densities on re-seeded grasslands than on control or fertilised grasslands. Another study from the UK found that geese grazed at higher densities on areas sown with clover, rather than grass seed. *Assessment: unknown effectiveness — limited evidence (effectiveness 35%; certainty 19%; harms 0%).*

http://www.conservationevidence.com/actions/352

Use environmentally sensitive flood management

One of two studies found more bird territories on a stretch of river in the UK with flood beams, compared to a channelized river. The other found that 13 out of 20 species of bird increased at sites in the USA where a river's hydrological dynamics were restored. *Assessment: unknown effectiveness — limited evidence (effectiveness 41%; certainty 26%; harms 0%).*

http://www.conservationevidence.com/actions/356

Use fire suppression/control

All three studies we captured, from the USA, UK and Australia, found that some bird species increased after fire suppression, and in one case that woodland species appeared in a site. Two studies (from the UK and USA) found that some species declined following fire suppression. The USA study identified open country species as being negatively affected. *Assessment: unknown effectiveness — limited evidence (effectiveness 35%; certainty 34%; harms 30%).*

http://www.conservationevidence.com/actions/324

Use greentree reservoir management

A study from the USA found that fewer mid- and under-storey birds were found at a greentree reservoir site than at a control site. Canopy-nesting species were not affected. *Assessment: unknown effectiveness — limited evidence (effectiveness 0%; certainty 10%; harms 40%).*

http://www.conservationevidence.com/actions/357

Use prescribed burning (Australian sclerophyll forest)

Two of three studies from Australia found no differences in bird species richness in burned sites compared to unburned areas. All three found differences in species assemblages, with some species lost and others gained from areas after fire. *Assessment: unknown effectiveness — limited evidence (effectiveness 30%; certainty 31%; harms 30%).*

http://www.conservationevidence.com/actions/319

Use shelterwood cutting instead of clearcutting

A study from the USA found that bird community composition differed between shelterwood stands and those under other forestry practices: some species were more abundant, others less so. *Assessment: unknown effectiveness — limited evidence (effectiveness 40%; certainty 20%; harms 40%).*

http://www.conservationevidence.com/actions/333

Use variable retention management during forestry operations

A study from the USA found that nine species were more abundant and five less so in stands under variable retention management, compared to unmanaged stands. *Assessment: unknown effectiveness — limited evidence (effectiveness 45%; certainty 20%; harms 25%).*

http://www.conservationevidence.com/actions/332

Likely to be ineffective or harmful

Apply herbicide to mid- and understorey vegetation

One of seven studies from North America found that bird species richness in a forest declined after deciduous trees were treated with herbicide. Three studies found increases in total bird densities, or those of some species, after herbicide treatment, although one found no differences between treatment and control areas. One study found that densities of one species decreased and another remained steady after treatment. Three studies found that nest survival was lower in herbicide-treated areas and one found lower nesting densities. One study found that northern bobwhite chicks higher had foraging success in forest areas treated with herbicide compared to under other managements. *Assessment: likely to be ineffective or harmful (effectiveness 20%; certainty 50%; harms 60%).*

http://www.conservationevidence.com/actions/346

Treat wetlands with herbicides

All four studies from the USA found higher densities of birds in wetlands sprayed with herbicide, compared with unsprayed areas. Two found that some species were at lower densities compared to unsprayed areas or those under other management. *Assessment: likely to be ineffective or harmful (effectiveness 30%; certainty 42%; harms 40%).*

http://www.conservationevidence.com/actions/347

Use prescribed burning (coastal habitats)

One study from the USA found that breeding seaside sparrow numbers decreased the year a site was burned, but were higher than on an unburned site the following year. One study in Argentina found that tall-grass specialist species were lost from burned areas in the year of burning, but that some habitats recovered by the following year. One study from the USA found no differences in nest predation rates between burned and unburned areas for two years after burning. *Assessment: likely to be ineffective or harmful (effectiveness 20%; certainty 40%; harms 30%).*

http://www.conservationevidence.com/actions/323

Use prescribed burning (deciduous forests)

One of four studies found that bird species richness was similar in burned and unburned aspen forests, although relative abundances of some species changed. A study in the USA found no changes in community composition in oak and hickory forests following burning. One study in the USA found no differences in wood thrush nest survival in burned and unburned areas. Another study in the USA found a reduction in black-chinned hummingbird nests following fuel reduction treatments including burning. *Assessments: likely to be ineffective or harmful (effectiveness 32%; certainty 60%; harms 30%).*

http://www.conservationevidence.com/actions/317

No evidence found (no assessment)

We have captured no evidence for the following interventions:

- Protect nest trees before burning

3.10 Habitat restoration and creation

Based on the collated evidence, what is the current assessment of the effectiveness of interventions for habitat restoration and creation?	
Beneficial	• Restore or create forests • Restore or create wetlands and marine habitats: restore or create inland wetlands
Likely to be beneficial	• Restore or create grassland • Restore or create traditional water meadows • Restore or create wetlands and marine habitats: restore or create coastal and intertidal wetlands
Unknown effectiveness (limited evidence)	• Restore or create shrubland • Restore or create wetlands and marine habitats: restore or create kelp forests • Restore or create wetlands and marine habitats: restore or create lagoons
No evidence found (no assessment)	• Restore or create savannahs • Revegetate gravel pits

Beneficial

Restore or create forests

Thirteen of 15 studies from across the world found that restored forests were similar to in-tact forests, that species returned to restored sites, that species recovered significantly better at restored than unrestored sites or that bird species richness, diversity or abundances in restored forest sites increased over time. One study also found that restoration techniques themselves improved over time. Nine studies found that some species did not return to restored forests or were less common and a study found that territory densities decreased over time. A study from the USA found that no more birds were found in restored sites, compared with unrestored. One study investigated productivity and found it was similar between restored and intact forests. A study from the USA found that planting fast-growing species appeared to provide better habitat than slower-growing trees. *Assessment: beneficial (effectiveness 65%; certainty 76%; harms 0%).*

http://www.conservationevidence.com/actions/360

● Restore or create wetlands and marine habitats (inland wetlands)

All eleven studies from the USA and Canada found that birds used restored or created wetlands. Two found that rates of use and species richness were similar or higher than on natural wetlands. One found that use was higher than on unrestored wetlands. Three studies from the USA and Puerto Rico found that restored wetlands held lower densities and fewer species or had similar productivity compared to natural wetlands. Two studies in the USA found that semi-permanent restored and larger wetlands were used more than temporary or seasonal or smaller ones. *Assessment: beneficial (effectiveness 70%; certainty 65%; harms 0%).*

http://www.conservationevidence.com/actions/366

Likely to be beneficial

● Restore or create grassland

Three of 23 studies found that species richness on restored grasslands was higher than unrestored habitats, or similar to remnant grassland, and three found that target species used restored grassland. Two studies from the USA found that diversity or species richness fell after restoration or was lower than unrestored sites. Seven studies from the USA and UK found high use of restored sites, or that such sites held a disproportionate proportion of the local population of birds. Two studies found that densities or abundances were lower on restored than unrestored sites, potentially due to drought conditions in one case. Five studies found that at least some bird species had higher productivities in restored sites compared to unrestored; had similar or higher productivities than natural habitats; or had high enough productivities to sustain populations. Three studies found that productivities were lower in restored than unrestored areas, or that productivities on restored sites were too low to sustain populations. A study from the USA found that older restored fields held more nests, but fewer species than young fields. Three studies found no differences between restoration techniques; two found that sowing certain species increased the use of sites by birds. *Assessment: likely to be beneficial (effectiveness 45%; certainty 70%; harms 0%).*

http://www.conservationevidence.com/actions/361

Restore or create traditional water meadows

Four out of five studies found that the number of waders or wildfowl on UK sites increased after the restoration of traditional water meadows. One study from Sweden found an increase in northern lapwing population after an increase in meadow management. One study found that lapwing productivity was higher on meadows than some habitats, but not others. *Assessment: likely to be beneficial (effectiveness 65%; certainty 50%; harms 0%).*

http://www.conservationevidence.com/actions/363

Restore or create wetlands and marine habitats (coastal and intertidal wetlands)

All six studies from the USA and UK found that bird species used restored or created wetlands. Two found that numbers and/or diversity were similar to in natural wetlands and one that numbers were higher than in unrestored sites. Three found that bird numbers on wetlands increased over time. Two studies from the UK found that songbirds and waders decreased following wetland restoration, whilst a study from the USA found that songbirds were more common on unrestored sites than restored wetlands. *Assessment: likely to be beneficial (effectiveness 65%; certainty 55%; harms 3%).*

http://www.conservationevidence.com/actions/367

Unknown effectiveness (limited evidence)

Restore or create shrubland

Three studies from the UK, USA and the Azores found local bird population increases after shrubland restoration. Two studies investigated multiple interventions and one found an increase from no birds to one or two pairs. One study from the UK found that several interventions, including shrubland restoration, were negatively related to the number of young grey partridges per adult bird on sites. *Assessment: unknown effectiveness — limited evidence (effectiveness 25%; certainty 20%; harms 3%).*

http://www.conservationevidence.com/actions/364

Restore or create wetlands and marine habitats (kelp forests)

One study in the USA found that the densities of five of the nine bird species increased following kelp forest restoration. *Assessment: unknown effectiveness — limited evidence (effectiveness 60%; certainty 15%; harms 0%).*

http://www.conservationevidence.com/actions/368

Restore or create wetlands and marine habitats (lagoons)

One study in the UK found that large numbers of bird species used and bred in a newly-created lagoon. *Assessment: unknown effectiveness — limited evidence (effectiveness 61%; certainty 20%; harms 0%).*

http://www.conservationevidence.com/actions/369

No evidence found (no assessment)

We have captured no evidence for the following interventions:

- Restore or create savannahs
- Revegetate gravel pits

3.11 Threat: Invasive alien and other problematic species

This assessment method for this chapter is described in Walsh, J. C., Dicks, L. V. & Sutherland, W. J. (2015) The effect of scientific evidence on conservation practitioners' management decisions. Conservation Biology, 29: 88–98. No harms were assessed for sections 3.11.1, 3.11.2, 3.11,3 and 3.11.4.

3.11.1 Reduce predation by other species

Based on the collated evidence, what is the current assessment of the effectiveness of interventions for reducing predation by other species?	
Beneficial	• Control mammalian predators on islands • Remove or control predators to enhance bird populations and communities
Likely to be beneficial	• Control avian predators on islands
Unknown effectiveness (limited evidence)	• Control invasive ants on islands • Reduce predation by translocating predators
Evidence not assessed	• Control predators not on islands

Beneficial

Control mammalian predators on islands

Of the 33 studies from across the world, 16 described population increases or recolonisations in at least some of the sites studied and 18 found higher reproductive success or lower mortality (on artificial nests in one case). Two studies that investigated population changes found only partial increases, in black oystercatchers *Haematopus bachmani* and two gamebird species, respectively. Eighteen of the studies investigated rodent control; 12 cat *Felis catus* control and 6 various other predators including pigs *Sus scrofa* and red foxes *Vulpes*. The two that found only partial increases examined cat, fox and other larger mammal removal. *Assessment: beneficial (effectiveness 81%; certainty 78%).*

http://www.conservationevidence.com/actions/373

Remove or control predators to enhance bird populations and communities

Both a meta-analysis and a systematic review (both global) found that bird reproductive success increased with predator control and that either post-breeding or breeding-season populations increased. The systematic review found that post-breeding success increased with predator control on mainland, but not islands. *Assessment: beneficial (effectiveness 66%; certainty 71%).*

http://www.conservationevidence.com/actions/371

Likely to be beneficial

Control avian predators on islands

Seven out of ten studies from North America, Australia and Europe found that controlling avian predators led to increased population sizes, reduced mortality, increased reproductive success or successful translocation of seabirds on islands. Two controlled studies on European islands found little effect of controlling crows on reproductive success in raptors or gamebirds. One study in the UK found that numbers of terns and small gulls on gravel islands declined despite the attempted control of large gulls. *Assessment: likely to be beneficial (effectiveness 50%; certainty 45%).*

http://www.conservationevidence.com/actions/372

Unknown effectiveness (limited evidence)

Control invasive ants on islands

A single study in the USA found that controlling the invasive tropical fire ant *Solenopsis geminata*, but not the big-headed ant *Pheidole megacephala*, led to lower rates of injuries and temporarily higher fledging success than on islands without ant control. The authors note that very few chicks were injured by *P. megacephala* on either experimental or control islands. *Assessment: unknown effectiveness — limited evidence (effectiveness 10%; certainty 15%).*

http://www.conservationevidence.com/actions/383

Reduce predation by translocating predators

Two studies from France and the USA found local population increases or reduced predation following the translocation of predators away from an area. *Assessment: unknown effectiveness — limited evidence (effectiveness 27%; certainty 20%).*

http://www.conservationevidence.com/actions/393

Evidence not assessed

Control predators not on islands

A study from the UK found higher bird community breeding densities and fledging success rates in plots with red fox *Vulpes vulpes* and carrion crow *Corvus corone* control. Of the 25 taxa-specific studies, only five found evidence for population increases with predator control, whilst one found a population decrease (with other interventions also used); one found lower or similar survival, probably because birds took bait. Nineteen studies found some evidence for increased reproductive success or decreased predation with predator control, with three studies (including a meta-analysis) finding no evidence for higher reproductive success or predation with predator control or translocation from the study site. One other study found evidence for increases in only three of six species studied. Most studies studied the removal of a number of different mammals, although several also removed bird predators, mostly carrion crows and gulls *Larus* spp. *Assessment: this intervention has not been assessed.*

http://www.conservationevidence.com/actions/384

3.11.2 Reduce incidental mortality during predator eradication or control

Based on the collated evidence, what is the current assessment of the effectiveness of interventions for reducing incidental mortality during predator eradication or control predation	
Unknown effectiveness (limited evidence)	• Distribute poison bait using dispensers • Use coloured baits to reduce accidental mortality during predator control • Use repellents on baits
Evidence not assessed	• Do birds take bait designed for pest control?

Unknown effectiveness (limited evidence)

Distribute poison bait using dispensers

A study from New Zealand found that South Island robin survival was higher when bait for rats and mice was dispensed from feeders, compared to being scattered. *Assessment: unknown effectiveness — limited evidence (effectiveness 40%; certainty 25%).*

http://www.conservationevidence.com/actions/157

Use coloured baits to reduce accidental mortality during predator control

Two out of three studies found that dyed baits were consumed at lower rates by songbirds and kestrels. An *ex situ* study from Australia found that dyeing food did not reduce its consumption by bush thick-knees. *Assessment: unknown effectiveness — limited evidence (effectiveness 20%; certainty 30%).*

http://www.conservationevidence.com/actions/182

Use repellents on baits

A study in New Zealand found that repellents reduced the rate of pecking at baits by North Island robins. A study from the USA found that treating bait with repellents did not reduce consumption by American kestrels. *Assessment: unknown effectiveness — limited evidence (effectiveness 10%; certainty 10%).*

http://www.conservationevidence.com/actions/159

Evidence not assessed

Do birds take bait designed for pest control?

Two studies from New Zealand and Australia, one *ex situ*, found no evidence that birds took bait meant for pest control. *Assessment: this intervention has not been assessed.*

http://www.conservationevidence.com/actions/395

3.11.3 Reduce nest predation by excluding predators from nests or nesting areas

Based on the collated evidence, what is the current assessment of the effectiveness of interventions for reducing nest predation by excluding predators from nests or nesting areas	
Likely to be beneficial	• Physically protect nests from predators using non-electric fencing • Physically protect nests with individual exclosures/barriers or provide shelters for chicks • Protect bird nests using electric fencing • Use artificial nests that discourage predation
Unknown effectiveness (limited evidence)	• Guard nests to prevent predation • Plant nesting cover to reduce nest predation • Protect nests from ants • Use multiple barriers to protect nests • Use naphthalene to deter mammalian predators • Use snakeskin to deter mammalian nest predators
No evidence found (no assessment)	• Play spoken-word radio programs to deter predators • Use 'cat curfews' to reduce predation • Use lion dung to deter domestic cats • Use mirrors to deter nest predators • Use ultrasonic devices to deter cats
Evidence not assessed	• Can nest protection increase nest abandonment? • Can nest protection increase predation of adults and chicks?

Likely to be beneficial

Physically protect nests from predators using non-electric fencing

Two of four studies from the UK and the USA found that fewer nests failed or were predated when predator exclusion fences were erected. Two studies found that nesting and fledging success was no higher when fences were used, one found that hatching success was higher. *Assessment: likely to be beneficial (effectiveness 45%; certainty 48%).*

http://www.conservationevidence.com/actions/183

Physically protect nests with individual exclosures/ barriers or provide shelters for chicks

Nine of 23 studies found that fledging rates or productivity were higher for nests protected by individual barriers than for unprotected nests. Two found no higher productivity. Fourteen studies found that hatching rates or survival were higher, or that predation was lower for protected nests. Two found no differences between protected and unprotected nests and one found that adults were harassed by predators at protected nests. One study found that chick shelters were not used much and a review found that some exclosure designs were more effective than others. *Assessment: likely to be beneficial (effectiveness 50%; certainty 50%).*

http://www.conservationevidence.com/actions/397
http://www.conservationevidence.com/actions/398
http://www.conservationevidence.com/actions/399
http://www.conservationevidence.com/actions/400

Protect bird nests using electric fencing

Two of six studies found increased numbers of terns or tern nests following the erection of an electric fence around colonies. Five studies found higher survival or productivity of waders or seabirds when electric fences were used and one found lower predation by mammals inside electric fences. One study found that predation by birds was higher inside electric fences. *Assessment: likely to be beneficial (effectiveness 60%; certainty 59%).*

http://www.conservationevidence.com/actions/188

Use artificial nests that discourage predation

Three out of five studies from North America found lower predation rates or higher nesting success for wildfowl in artificial nests, compared with natural nests. An *ex situ* study found that some nest box designs prevented raccoons from entering. A study found that wood ducks avoided anti-predator nest boxes but only if given the choice of unaltered nest boxes. *Assessment: likely to be beneficial (effectiveness 59%; certainty 54%).*

http://www.conservationevidence.com/actions/402

Unknown effectiveness (limited evidence)

Guard nests to prevent predation

Nest guarding can be used as a response to a range of threats and is therefore discussed in 'General responses to small/declining populations — Guard nests'. *Assessment: unknown effectiveness — limited evidence (effectiveness 50%; certainty 30%).*

http://www.conservationevidence.com/actions/411

Plant nesting cover to reduce nest predation

Studies relevant to this intervention are discussed in 'Threat: Agriculture'. *Assessment: unknown effectiveness — limited evidence (effectiveness 28%; certainty 30%).*

http://www.conservationevidence.com/actions/405

Protect nests from ants

A study from the USA found that vireo nests protected from ants with a physical barrier and a chemical repellent had higher fledging success than unprotected nests. *Assessment: unknown effectiveness — limited evidence (effectiveness 45%; certainty 17%).*

http://www.conservationevidence.com/actions/410

Use multiple barriers to protect nests

One of two studies found that plover fledging success in the USA was no higher when an electric fence was erected around individual nest exclosures, compared to when just the exclosures were present. A study from the USA found that predation on chicks was lower when one of two barriers around nests was removed early, compared to when it was left for three more days. *Assessment: unknown effectiveness — limited evidence (effectiveness 7%; certainty 17%).*

http://www.conservationevidence.com/actions/404

Use naphthalene to deter mammalian predators

A study from the USA found that predation rates on artificial nests did not differ when naphthalene moth balls were scattered around them. *Assessment: unknown effectiveness — limited evidence (effectiveness 0%; certainty 10%).*

http://www.conservationevidence.com/actions/408

Use snakeskin to deter mammalian nest predators

A study from the USA found that flycatcher nests were predated less frequently if they had a snakeskin wrapped around them. *Assessment: unknown effectiveness — limited evidence (effectiveness 33%; certainty 15%).*

http://www.conservationevidence.com/actions/406

No evidence found (no assessment)

We have captured no evidence for the following interventions:

- Play spoken-word radio programmes to deter predators
- Use 'cat curfews' to reduce predation
- Use lion dung to deter domestic cats
- Use mirrors to deter nest predators
- Use ultrasonic devices to deter cats

Evidence not assessed

Can nest protection increase nest abandonment?

One of four studies (from the USA) found an increase in abandonment after nest exclosures were used. Two studies from the USA and Sweden found no increases in abandonment when exclosures were used and a review from the USA found that some designs were more likely to cause abandonment than others. *Assessment: this intervention has not been assessed.*

http://www.conservationevidence.com/actions/401

Can nest protection increase predation of adults and chicks?

Four of five studies from the USA and Sweden found that predation on chicks and adults was higher when exclosures were used. One of these found that adults were harassed when exclosures were installed and the chicks rapidly predated when they were removed. One study from Sweden found that predation was no higher when exclosures were used. *Assessment: this intervention has not been assessed.*

http://www.conservationevidence.com/actions/403

3.11.4 Reduce mortality by reducing hunting ability or changing predator behaviour

Based on the collated evidence, what is the current assessment of the effectiveness of interventions for reducing mortality by reducing hunting ability or changing predator behaviour	
Unknown effectiveness (limited evidence)	• Reduce predation by translocating nest boxes • Use collar-mounted devices to reduce predation • Use supplementary feeding of predators to reduce predation
Unlikely to be beneficial	• Use aversive conditioning to reduce nest predation

Unknown effectiveness (limited evidence)

Reduce predation by translocating nest boxes

Two European studies found that predation rates were lower for translocated nest boxes than for controls. *Assessment: unknown effectiveness — limited evidence (effectiveness 48%; certainty 25%).*

http://www.conservationevidence.com/actions/420

Use collar-mounted devices to reduce predation

Two replicated randomised and controlled studies in the UK and Australia found that fewer birds were returned by cats wearing collars with anti-hunting devices, compared to cats with control collars. No differences were found between different devices. *Assessment: unknown effectiveness — limited evidence (effectiveness 48%; certainty 35%).*

http://www.conservationevidence.com/actions/416

Use supplementary feeding to reduce predation

One of three studies found that fewer grouse chicks were taken to harrier nests when supplementary food was provided to the harriers, but no effect on grouse adult survival or productivity was found. One study from the USA found reduced predation on artificial nests when supplementary food was provided. Another study from the USA found no such effect. *Assessment: unknown effectiveness — limited evidence (effectiveness 13%; certainty 20%).*

http://www.conservationevidence.com/actions/417

Unlikely to be beneficial

Use aversive conditioning to reduce nest predation

Nine out of 12 studies found no evidence for aversive conditioning or reduced nest predation after aversive conditioning treatment stopped. Ten studies found reduced consumption of food when it was treated with repellent chemicals, i.e. during the treatment. Three, all studying avian predators, found some evidence for reduced consumption after treatment but these were short-lived trials or the effect disappeared within a year. *Assessment: unlikely to be beneficial (effectiveness 9%; certainty 60%).*

http://www.conservationevidence.com/actions/418

http://www.conservationevidence.com/actions/419

3.11.5 Reduce competition with other species for food and nest sites

Based on the collated evidence, what is the current assessment of the effectiveness of interventions for reducing competition with other species for food and nest sites?	
Likely to be beneficial	• Reduce inter-specific competition for food by removing or controlling competitor species
Unknown effectiveness (limited evidence)	• Protect nest sites from competitors • Reduce competition between species by providing nest boxes • Reduce inter-specific competition for nest sites by modifying habitats to exclude competitor species • Reduce inter-specific competition for nest sites by removing competitor species: ground nesting seabirds • Reduce inter-specific competition for nest sites by removing competitor species: songbirds • Reduce inter-specific competition for nest sites by removing competitor species: woodpeckers

Likely to be beneficial

Reduce inter-specific competition for food by removing or controlling competitor species

Three out of four studies found that at least some of the target species increased following the removal or control of competitor species. Two studies found that some or all target species did not increase, or that there was no change in kleptoparasitic behaviour of competitor species after control efforts. *Assessment: likely to be beneficial (effectiveness 44%; certainty 40%; harms 0%).*

http://www.conservationevidence.com/actions/428

Unknown effectiveness (limited evidence)

Protect nest sites from competitors

Two studies from the USA found that red-cockaded woodpecker populations increased after the installation of 'restrictor plates' around nest holes to prevent larger woodpeckers for enlarging them. Several other interventions were used at the same time. A study from Puerto Rico found lower competition between species after nest boxes were altered. A study from the USA found weak evidence that exclusion devices prevented house sparrows from using nest boxes and another study from the USA found that fitting restrictor plates to red-cockaded woodpecker holes reduced the number that were enlarged by other woodpeckers. *Assessment: unknown effectiveness — limited evidence (effectiveness 39%; certainty 24%; harms 5%).*

http://www.conservationevidence.com/actions/426

Reduce competition between species by providing nest boxes

A study from the USA found that providing extra nest boxes did not reduce the rate at which common starlings usurped northern flickers from nests. *Assessment: unknown effectiveness — limited evidence (effectiveness 0%; certainty 16%; harms 0%).*

http://www.conservationevidence.com/actions/427

Reduce inter-specific competition for nest sites by modifying habitats to exclude competitor species

A study from the USA found that clearing midstorey vegetation did not reduce the occupancy of red-cockaded woodpecker nesting holes by southern flying squirrels. *Assessment: unknown effectiveness — limited evidence (effectiveness 0%; certainty 12%; harms 0%).*

http://www.conservationevidence.com/actions/425

Reduce inter-specific competition for nest sites by removing competitor species (ground nesting seabirds)

Four studies from Canada and the UK found increased tern populations following the control or exclusion of gulls, and in two cases with many additional interventions. Two studies from the UK and Canada found that controlling large gulls had no impact on smaller species. Two studies from the USA and UK found that exclusion devices successfully reduced the numbers of gulls at sites, although one found that they were only effective at small colonies and the other found that methods varied in their effectiveness and practicality. *Assessment: unknown effectiveness — limited evidence (effectiveness 41%; certainty 31%; harms 14%).*

http://www.conservationevidence.com/actions/422

Reduce inter-specific competition for nest sites by removing competitor species (songbirds)

Two studies from Australia found increases in bird populations and species richness after control of noisy miners. A study from Italy found that blue tits nested in more nest boxes when hazel dormice were excluded from boxes over winter. *Assessment: unknown effectiveness — limited evidence (effectiveness 50%; certainty 22%; harms 0%).*

http://www.conservationevidence.com/actions/424

Reduce inter-specific competition for nest sites by removing competitor species (woodpeckers)

Two studies in the USA found red-cockaded woodpecker populations increased following the removal of southern flying squirrels, in one case along with other interventions. A third found that red-cockaded woodpecker reintroductions were successful when squirrels were controlled. One study found fewer holes were occupied by squirrels following control efforts, but that occupancy by red-cockaded woodpeckers was no higher. *Assessment: unknown effectiveness — limited evidence (effectiveness 34%; certainty 28%; harms 0%).*

http://www.conservationevidence.com/actions/423

3.11.6 Reduce adverse habitat alteration by other species

Based on the collated evidence, what is the current assessment of the effectiveness of interventions for reducing adverse habitat alteration by other species?	
Likely to be beneficial	• Control or remove habitat-altering mammals • Reduce adverse habitat alterations by excluding problematic species (terrestrial species)
Unknown effectiveness (limited evidence)	• Reduce adverse habitat alterations by excluding problematic species (aquatic species) • Remove problematic vegetation • Use buffer zones to reduce the impact of invasive plant control

Likely to be beneficial

Control or remove habitat-altering mammals

Four out of five studies from islands in the Azores and Australia found that seabird populations increased after rabbits or other species were removed, although three studied several interventions at the same time. Two studies from Australia and Madeira found that seabird productivity increased after rabbit and house mouse eradication. *Assessment: likely to be beneficial (effectiveness 61%; certainty 41%; harms 0%).*

http://www.conservationevidence.com/actions/431

Reduce adverse habitat alterations by excluding problematic species (terrestrial species)

Three studies from the USA and the UK found higher numbers of certain songbird species and higher species richness in these groups when deer were excluded from forests. Intermediate canopy-nesting species in the USA and common nightingales in the UK were the species to benefit. A study from Hawaii found mixed effects of grazer exclusion. *Assessment: likely to be beneficial (effectiveness 48%; certainty 40%; harms 0%).*

http://www.conservationevidence.com/actions/429

Unknown effectiveness (limited evidence)

Reduce adverse habitat alterations by excluding problematic species (aquatic species)

A study in the USA found that waterbirds preferentially used wetland plots from which grass carp were excluded but moved as these became depleted over the winter. *Assessment: unknown effectiveness — limited evidence (effectiveness 30%; certainty 14%; harms 0%).*

http://www.conservationevidence.com/actions/430

Remove problematic vegetation

One of four studies (from Japan) found an increase in a bird population following the removal of an invasive plant. One study from the USA found lower bird densities in areas where a problematic native species was removed. One study from Australia found the Gould's petrel productivity was higher following the removal of native bird-lime trees, and a study from New Zealand found that Chatham Island oystercatchers could nest in preferable areas of beaches after invasive marram grass was removed. *Assessment: unknown effectiveness — limited evidence (effectiveness 43%; certainty 23%; harms 0%).*

http://www.conservationevidence.com/actions/432

Use buffer zones to reduce the impact of invasive plant control

A study from the USA found that no snail kite nests (built above water in cattail and bulrush) were lost during herbicide spraying when buffer zones were established around nests. *Assessment: unknown effectiveness — limited evidence (effectiveness 40%; certainty 10%; harms 0%).*

http://www.conservationevidence.com/actions/433

3.11.7 Reduce parasitism and disease

Based on the collated evidence, what is the current assessment of the effectiveness of interventions for reducing parasitism and disease?	
Likely to be beneficial	• Remove/control adult brood parasites
Trade-off between benefit and harms	• Remove/treat endoparasites and diseases
Unknown effectiveness (limited evidence)	• Alter artificial nest sites to discourage brood parasitism • Exclude or control 'reservoir species' to reduce parasite burdens • Remove brood parasite eggs from target species' nests • Remove/treat ectoparasites to increase survival or reproductive success: reduce nest ectoparasites by providing beneficial nesting material • Remove/treat ectoparasites to increase survival or reproductive success: remove ectoparasites from feathers • Use false brood parasite eggs to discourage brood parasitism
Unlikely to be beneficial	• Remove/treat ectoparasites to increase survival or reproductive success: remove ectoparasites from nests

Likely to be beneficial

Remove/control adult brood parasites

One of 12 studies, all from the Americas, found that a host species population increased after control of the parasitic cowbird, two studies found no effect. Five studies found higher productivities or success rates when cowbirds were removed, five found that some or all measures of productivity were no different. Eleven studies found that brood parasitism rates were lower after cowbird control. *Assessment: likely to be beneficial (effectiveness 48%; certainty 61%; harms 0%).*

http://www.conservationevidence.com/actions/441

Trade-off between benefit and harms

Remove/treat endoparasites and diseases

Two out of five studies found that removing endoparasites increased survival in birds and one study found higher productivity in treated birds. Two studies found no evidence, or uncertain evidence, for increases in survival with treatment and one study found lower parasite burdens, but also lower survival in birds treated with antihelmintic drugs. *Assessment: trade-offs between benefits and harms (effectiveness 48%; certainty 51%; harms 37%).*

http://www.conservationevidence.com/actions/434

Unknown effectiveness (limited evidence)

Alter artificial nest sites to discourage brood parasitism

A replicated trial from Puerto Rico found that brood parasitism levels were extremely high across all nest box designs tested. *Assessment: unknown effectiveness — limited evidence (effectiveness 0%; certainty 13%; harms 0%).*

http://www.conservationevidence.com/actions/446

Exclude or control 'reservoir species' to reduce parasite burdens

One of two studies found increased chick production in grouse when hares (carries of louping ill virus) were culled in the area, although a comment on the paper disputes this finding. A literature review found no compelling evidence for the effects of hare culling on grouse populations. *Assessment: unknown effectiveness — limited evidence (effectiveness 13%; certainty 20%; harms 0%).*

http://www.conservationevidence.com/actions/435

Remove brood parasite eggs from target species' nests

One of two studies found lower rates of parasitism when cowbird eggs were removed from host nests. One study found that nests from which cowbird eggs were removed had lower success than parasitised nests. *Assessment: unknown effectiveness — limited evidence (effectiveness 24%; certainty 20%; harms 21%).*

http://www.conservationevidence.com/actions/443

Remove/treat ectoparasites to increase survival or reproductive success (provide beneficial nesting material)

A study in Canada found lower numbers of some, but not all, parasites in nests provided with beneficial nesting material, but that there was no effect on fledging rates or chick condition. *Assessment: unknown effectiveness — limited evidence (effectiveness 15%; certainty 13%; harms 0%).*

http://www.conservationevidence.com/actions/439

Remove/treat ectoparasites to increase survival or reproductive success (remove ectoparasites from feathers)

A study in the UK found that red grouse treated with spot applications had lower tick and disease burdens and higher survival than controls, whilst birds with impregnated tags had lower tick burdens only. A study in Hawaii found that CO2 was the most effective way to remove lice from feathers, although lice were not killed. *Assessment: unknown effectiveness — limited evidence (effectiveness 42%; certainty 16%; harms 0%).*

http://www.conservationevidence.com/actions/437

Use false brood parasite eggs to discourage brood parasitism

A study from the USA found that parasitism rates were lower for red-winged blackbird nests with false or real cowbird eggs placed in them, than for control nests. *Assessment: unknown effectiveness — limited evidence (effectiveness 35%; certainty 19%; harms 0%).*

http://www.conservationevidence.com/actions/444

Unlikely to be beneficial

Remove/treat ectoparasites to increase survival or reproductive success (remove ectoparasites from nests)

Six of the seven studies found lower infestation rates in nests treated for ectoparasites, one (that used microwaves to treat nests) did not find fewer parasites. Two studies from the USA found higher survival or lower abandonment in nests treated for ectoparasites, whilst seven studies from across the world found no differences in survival, fledging rates or productivity between nests treated for ectoparasites and controls. Two of six studies found that chicks from nests treated for ectoparasites were in better condition than those from control nests. *Assessment: unlikely to be beneficial (effectiveness 25%; certainty 58%; harms 0%).*

http://www.conservationevidence.com/actions/438

3.11.8 Reduce detrimental impacts of other problematic species

Based on the collated evidence, what is the current assessment of the effectiveness of interventions for reducing detrimental impacts of other problematic species?	
Unknown effectiveness (limited evidence)	• Use copper strips to exclude snails from nests

Unknown effectiveness (limited evidence)

Use copper strips to exclude snails from nests

A study from Mauritius found no mortality from snails invading echo parakeet nests after the installation of copper strips around nest trees. Before installation, four chicks were killed by snails. *Assessment: unknown effectiveness — limited evidence (effectiveness 47%; certainty 15%; harms 0%).*

http://www.conservationevidence.com/actions/447

3.12 Threat: Pollution

3.12.1 Industrial pollution

Based on the collated evidence, what is the current assessment of the effectiveness of interventions for industrial pollution?	
Likely to be beneficial	• Use visual and acoustic 'scarers' to deter birds from landing on pools polluted by mining or sewage
Unknown effectiveness (limited evidence)	• Relocate birds following oil spills • Use repellents to deter birds from landing on pools polluted by mining
Unlikely to be beneficial	• Clean birds after oil spills

Likely to be beneficial

Use visual and acoustic 'scarers' to deter birds from landing on pools polluted by mining or sewage

Two studies from Australia and the USA found that deterrent systems reduced bird mortality on toxic pools. Four of five studies from the USA and Canada found that fewer birds landed on pools when deterrents were used, one found no effect. Two studies found that radar-activated systems were more effective than randomly-activated systems. One study found that loud noises were more effective than raptor models. *Assessment: likely to be beneficial (effectiveness 50%; certainty 46%; harms 0%).*

http://www.conservationevidence.com/actions/452

Unknown effectiveness (limited evidence)

Relocate birds following oil spills

A study from South Africa found that a high percentage of penguins relocated following an oil spill returned to and bred at their old colony. More relocated birds bred than oiled-and-cleaned birds. *Assessment: unknown effectiveness — limited evidence (effectiveness 39%; certainty 10%; harms 5%).*

http://www.conservationevidence.com/actions/449

Use repellents to deter birds from landing on pools polluted by mining

An *ex situ* study from the USA found that fewer common starlings consumed contaminated water laced with chemicals, compared to untreated water. *Assessment: unknown effectiveness — limited evidence (effectiveness 51%; certainty 10%; harms 0%).*

http://www.conservationevidence.com/actions/453

Unlikely to be beneficial

Clean birds after oil spills

Three studies from South Africa and Australia found high survival of oiled-and-cleaned penguins and plovers, but a large study from the USA found low survival of cleaned common guillemots. Two studies found that cleaned birds bred and had similar success to un-oiled birds. After a second spill, one study found that cleaned birds were less likely to breed. Two studies found that cleaned birds had lower breeding success than un-oiled birds. *Assessment: unlikely to be beneficial (effectiveness 30%; certainty 45%; harms 5%).*

http://www.conservationevidence.com/actions/448

3.12.2 Agricultural pollution

Based on the collated evidence, what is the current assessment of the effectiveness of interventions for agricultural pollution?	
Likely to be beneficial	• Leave headlands in fields unsprayed (conservation headlands) • Provide food for vultures to reduce mortality from diclofenac • Reduce pesticide, herbicide and fertiliser use generally
Unknown effectiveness (limited evidence)	• Reduce chemical inputs in permanent grassland management • Restrict certain pesticides or other agricultural chemicals
No evidence found (no assessment)	• Make selective use of spring herbicides • Provide buffer strips along rivers and streams • Provide unfertilised cereal headlands in arable fields • Use buffer strips around in-field ponds • Use organic rather than mineral fertilisers

Likely to be beneficial

Leave headlands in fields unsprayed (conservation headlands)

Three studies from Europe found that several species were strongly associated with conservation headlands; two of these found that other species were not associated with them. A review from the UK found larger grey partridge populations on sites with conservation headlands. Three studies found higher grey partridge adult or chick survival on sites with conservation headlands, one found survival did not differ. Four studies found higher grey partridge productivity on sites with conservation headlands, two found similar productivities and one found a negative relationship between conservation headlands and the number of chicks per adult partridge. *Assessment: likely to be beneficial (effectiveness 70%; certainty 50%; harms 0%).*

http://www.conservationevidence.com/actions/461

Provide food for vultures to reduce mortality from diclofenac

A before-and-after trial in Pakistan found that oriental white-backed vulture mortality rates were significantly lower when supplementary food was provided, compared to when it was not. *Assessment: likely to be beneficial (effectiveness 60%; certainty 40%; harms 0%).*

http://www.conservationevidence.com/actions/456

Reduce pesticide, herbicide and fertiliser use generally

One of nine studies found that the populations of some species increased when pesticide use was reduced and other interventions used. Three studies found that some or all species were found at higher densities on reduced-input sites. Five found that some of all species were not at higher densities. A study from the UK found that grey partridge chicks had higher survival on sites with reduced pesticide input. Another found that partridge broods were smaller on such sites and there was no relationship between reduced inputs and survival or the ratio of young to old birds. *Assessment: likely to be beneficial (effectiveness 55%; certainty 55%; harms 3%).*

http://www.conservationevidence.com/actions/454

Unknown effectiveness (limited evidence)

Reduce chemical inputs in permanent grassland management

A study from the UK found that no more foraging birds were attracted to pasture plots with no fertiliser, compared to control plots. *Assessment: unknown effectiveness — limited evidence (effectiveness 0%; certainty 10%; harms 0%).*

http://www.conservationevidence.com/actions/459

Restrict certain pesticides or other agricultural chemicals

A before-and-study from Spain found an increase in the regional griffon vulture population following the banning of strychnine, amongst several other interventions. *Assessment: unknown effectiveness — limited evidence (effectiveness 20%; certainty 10%; harms 0%).*

http://www.conservationevidence.com/actions/455

No evidence found (no assessment)

We have captured no evidence for the following interventions:

- Make selective use of spring herbicides
- Provide buffer strips along rivers and streams
- Provide unfertilised cereal headlands in arable fields
- Use buffer strips around in-field ponds
- Use organic rather than mineral fertilisers

3.12.3 Air-borne pollutants

Based on the collated evidence, what is the current assessment of the effectiveness of interventions for air-borne pollutants?	
Unknown effectiveness (limited evidence)	• Use lime to reduce acidification in lakes

Unknown effectiveness (limited evidence)

Use lime to reduce acidification in lakes

A study from Sweden found no difference in osprey productivity during a period of extensive liming of acidified lakes compared to two periods without liming. *Assessment: unknown effectiveness — limited evidence (effectiveness 0%; certainty 10%; harms 0%).*

http://www.conservationevidence.com/actions/465

3.12.4 Excess energy

Based on the collated evidence, what is the current assessment of the effectiveness of interventions for excess energy?	
Unknown effectiveness (limited evidence)	• Shield lights to reduce mortality from artificial lights • Turning off lights to reduce mortality from artificial lights • Use flashing lights to reduce mortality from artificial lights • Use lights low in spectral red to reduce mortality from artificial lights
No evidence found (no assessment)	• Reduce the intensity of lighthouse beams • Using volunteers to collect and rehabilitate downed birds

Unknown effectiveness (limited evidence)

Shield lights to reduce mortality from artificial lights

A study from the USA found that fewer shearwaters were downed when security lights were shielded, compared to nights with unshielded lights. *Assessment: unknown effectiveness — limited evidence (effectiveness 50%; certainty 15%; harms 0%).*

http://www.conservationevidence.com/actions/469

Turning off lights to reduce mortality from artificial lights

A study from the UK found that fewer seabirds were downed when artificial (indoor and outdoor) lighting was reduced at night, compared to nights with normal lighting. *Assessment: unknown effectiveness — limited evidence (effectiveness 49%; certainty 10%; harms 0%).*

http://www.conservationevidence.com/actions/467

Use flashing lights to reduce mortality from artificial lights

A study from the USA found that fewer dead birds were found beneath aviation control towers with only flashing lights, compared to those with both flashing and continuous lights. *Assessment: unknown effectiveness — limited evidence (effectiveness 54%; certainty 15%; harms 0%).*

http://www.conservationevidence.com/actions/470

Use lights low in spectral red to reduce mortality from artificial lights

Two studies from Europe found that fewer birds were attracted to low-red lights (including green and blue lights), compared with the number expected, or the number attracted to white or red lights. *Assessment: unknown effectiveness — limited evidence (effectiveness 56%; certainty 15%; harms 0%).*

http://www.conservationevidence.com/actions/471

No evidence found (no assessment)

We have captured no evidence for the following interventions:

- Reduce the intensity of lighthouse beams
- Using volunteers to collect and rehabilitate downed birds

3.13 Threat: Climate change, extreme weather and geological events

Based on the collated evidence, what is the current assessment of the effectiveness of interventions for climate change, extreme weather and geological events?	
Unknown effectiveness (limited evidence)	• Replace nesting habitats when they are washed away by storms • Water nesting mounds to increase incubation success in malleefowl

Unknown effectiveness (limited evidence)

Replace nesting habitats when they are washed away by storms

A before-and-after study found that a common tern colony increased following the replacement of nesting habitats, whilst a second found that a colony decreased. In both cases, several other interventions were used at the same time, making it hard to examine the effect of habitat provision. *Assessment: unknown effectiveness — limited evidence (effectiveness 8%; certainty 10%; harms 0%).*

http://www.conservationevidence.com/actions/474

Water nesting mounds to increase incubation success in malleefowl

A single small trial in Australia found that watering malleefowl nests increased their internal temperature but that a single application of water did not prevent the nests drying out and being abandoned during a drought. *Assessment: unknown effectiveness — limited evidence (effectiveness 9%; certainty 10%; harms 0%).*

http://www.conservationevidence.com/actions/473

3.14 General responses to small/declining populations

3.14.1 Inducing breeding, rehabilitation and egg removal

Based on the collated evidence, what is the current assessment of the effectiveness of interventions for inducing breeding, rehabilitation and egg removal?	
Unknown effectiveness (limited evidence)	• Rehabilitate injured birds • Remove eggs from wild nests to increase reproductive output • Use artificial visual and auditory stimuli to induce breeding in wild populations

Unknown effectiveness (limited evidence)

Rehabilitate injured birds

Two studies of four studies from the UK and USA found that 25–40% of injured birds taken in by centres were rehabilitated and released. Three studies from the USA found that rehabilitated birds appeared to have high survival. One found that mortality rates were higher for owls than raptors. *Assessment: unknown effectiveness — limited evidence (effectiveness 36%; certainty 30%; harms 0%).*

http://www.conservationevidence.com/actions/476

Remove eggs from wild nests to increase reproductive output

A study from Canada found that whooping crane reproductive success was higher for nests with one or two eggs removed than for controls. A study from the USA found that removing bald eagle eggs did not appear to affect the wild population and a replicated study from Mauritius found that removing entire Mauritius kestrel clutches appeared to increase productivity more than removing individual eggs as they were laid. *Assessment: unknown effectiveness — limited evidence (effectiveness 24%; certainty 25%; harms 5%).*

http://www.conservationevidence.com/actions/477

Use artificial visual and auditory stimuli to induce breeding in wild populations

A small study from the British Virgin Islands found an increase in breeding behaviour after the introduction of visual and auditory stimulants. *Assessment: unknown effectiveness — limited evidence (effectiveness 19%; certainty 11%; harms 0%).*

http://www.conservationevidence.com/actions/475

3.14.2 Provide artificial nesting sites

Based on the collated evidence, what is the current assessment of the effectiveness of interventions for providing artificial nesting sites?	
Beneficial	• Provide artificial nests: falcons • Provide artificial nests: owls • Provide artificial nests: songbirds • Provide artificial nests: wildfowl
Likely to be beneficial	• Clean artificial nests to increase occupancy or reproductive success • Provide artificial nests: burrow-nesting seabirds • Provide artificial nests: divers/loons • Provide artificial nests: ground- and tree-nesting seabirds • Provide artificial nests: oilbirds • Provide artificial nests: raptors • Provide artificial nests: wildfowl — artificial/ floating islands

Unknown effectiveness (limited evidence)	• Artificially incubate eggs or warm nests • Guard nests • Provide artificial nests: gamebirds • Provide artificial nests: grebes • Provide artificial nests: ibises and flamingos • Provide artificial nests: parrots • Provide artificial nests: pigeons • Provide artificial nests: rails • Provide artificial nests: rollers • Provide artificial nests: swifts • Provide artificial nests: trogons • Provide artificial nests: waders • Provide artificial nests: woodpeckers • Provide nesting habitat for birds that is safe from extreme weather • Provide nesting material for wild birds • Remove vegetation to create nesting areas • Repair/support nests to support breeding • Use differently-coloured artificial nests

Beneficial

● Provide artificial nests (falcons)

Four studies from the USA and Europe found that local populations of falcons increased following the installation of artificial nesting sites. However, a study from Canada found no increase in the local population of falcons following the erection of nest boxes. Eight studies from across the world found that the success and productivity of falcons in nest boxes was higher than or equal to those in natural nests. Four studies from across the world found that productivity in nest boxes was lower than in natural nests, or that some falcons were evicted from their nests by owls. Four studies from across the world found no differences in productivity between nest box designs or positions, whilst two from Spain and Israel found that productivity in boxes varied between designs and habitats. Twenty-one studies from across the world found nest boxes were used by falcons, with one in the UK finding that nest boxes were not used at all. Seven studies found that position or design affected use, whilst three found no differences between design or positioning. *Assessment: beneficial (effectiveness 65%; certainty 65%; harms 0%).*

http://www.conservationevidence.com/actions/489

Provide artificial nests (owls)

Three studies from the UK appeared to show increases in local populations of owls following the installation of artificial nests. Another UK study found that providing nesting sites when renovating buildings maintained owl populations, whilst they declined at sites without nests. Four studies from the USA and the UK found high levels of breeding success in artificial nests. Two studies from the USA and Hungary found lower productivity or fledgling survival from breeding attempts in artificial nests, whilst a study from Finland found that artificial nests were only successful in the absence of larger owls. Four studies from the USA and Europe found that artificial nests were used as frequently as natural sites. Five studies from across the world found that owls used artificial nests. Seven studies found that nest position or design affected occupancy or productivity. However four studies found occupancy and/or productivity did not differ between different designs of nest box. *Assessment: beneficial (effectiveness 65%; certainty 66%; harms 5%).*

http://www.conservationevidence.com/actions/490

Provide artificial nests (songbirds)

Only three out of 66 studies from across the world found low rates of nest box occupancy in songbirds. Low rates of use were seen in thrushes, crows, swallows and New World warblers. Thrushes, crows, finches, swallows, wrens, tits, Old World and tyrant flycatchers, New World blackbirds, sparrows, waxbills, starlings and ovenbirds all used nest boxes. Five studies from across the world found higher population densities or growth rates, and one study from the USA found higher species richness, in areas with nest boxes. Twelve studies from across the world found that productivity in nest boxes was higher than or similar to natural nests. One study found there were more nesting attempts in areas with more nest boxes, although a study from Canada found no differences in productivity between areas with different nest box densities. Two studies from Europe found lower predation of species using nest boxes but three studies from the USA found low production in nest boxes. Thirteen studies from across the world found that use, productivity or usurpation rate varied with nest box design, whilst seven found no difference in occupation rates or success between different designs. Similarly, fourteen studies found different occupation or success rates depending on the position of artificial nest sites but two studies found no such differences. *Assessment: beneficial (effectiveness 67%; certainty 85%; harms 0%).*

http://www.conservationevidence.com/actions/498

● Provide artificial nests (wildfowl)

Six studies from North America and Europe found that wildfowl populations increased with the provision of artificial nests, although one study from Finland found no increase in productivity in areas with nest boxes. Nine out of twelve studies from North America found that productivity was high in artificial nests. Two studies found that success for some species in nest boxes was lower than for natural nests. Nineteen studies from across the world found that occupancy rates varied from no use to 100% occupancy. Two studies found that occupancy rates were affected by design or positioning. Three studies from North America found that nest boxes could have other impacts on reproduction and behaviour. *Assessment: beneficial (effectiveness 62%; certainty 76%; harms 0%).*

http://www.conservationevidence.com/actions/482

Likely to be beneficial

● Clean artificial nests to increase occupancy or reproductive success

Five out of ten studies from North America and Europe found that songbirds preferentially nested in cleaned nest boxes or those sterilised using microwaves, compared to used nest boxes. One study found that the preference was not strong enough for birds to switch nest boxes after they were settled. One study found that birds avoided heavily-soiled nest boxes. Two studies birds had a preference for used nest boxes and one found no preference for cleaned or uncleaned boxes. None of the five studies that examined it found any effect of nest box cleanliness on nesting success or parasitism levels. *Assessment: likely to be beneficial (effectiveness 40%; certainty 40%; harms 15%).*

http://www.conservationevidence.com/actions/499

Provide artificial nests (burrow-nesting seabirds)

Four studies from across the world found population increases or population establishment following the provision of nest boxes. In two cases this was combined with other interventions. Six studies from across the world found high occupancy rates for artificial burros by seabirds but three studies from across the world found very low occupancy rates for artificial burrows used by petrels. Eight studies from across the world found that the productivity of birds in artificial burrows was high although two studies from the USA and the Galapagos found low productivity in petrels. *Assessment: likely to be beneficial (effectiveness 60%; certainty 71%; harms 5%).*

http://www.conservationevidence.com/actions/481

Provide artificial nests (divers/loons)

Three studies from the UK and the USA found increases in loon productivity on lakes provided with nesting rafts. A study in the UK found that usage of nesting rafts varied between sites. *Assessment: likely to be beneficial (effectiveness 50%; certainty 50%; harms 0%).*

http://www.conservationevidence.com/actions/478

Provide artificial nests (ground- and tree-nesting seabirds)

Three studies from the UK and the Azores found increases in gull and tern populations following the provision of rafts/islands or nest boxes alongside other interventions. Five studies from Canada and Europe found that terns used artificial nesting sites. A study from the USA found that terns had higher nesting success on artificial rafts in some years and a study from Japan found increased nesting success after provision of nesting substrate. Design of nesting structure should be considered. *Assessment: likely to be beneficial (effectiveness 60%; certainty 49%; harms 0%).*

http://www.conservationevidence.com/actions/480

Provide artificial nests (oilbirds)

A study in Trinidad and Tobago found an increase in the size of an oilbird colony after the creation of artificial nesting lodges. *Assessment: likely to be beneficial (effectiveness 50%; certainty 45%; harms 0%).*

http://www.conservationevidence.com/actions/491

Provide artificial nests (raptors)

Nine studies from North America and Spain found that raptors used artificial nesting platforms. Two studies from the USA found increases in populations or densities following the installation of platforms. Three studies describe successful use of platforms but three found lower productivity or failed nesting attempts, although these studies only describe a single nesting attempt. *Assessment: likely to be beneficial (effectiveness 55%; certainty 55%; harms 0%).*

http://www.conservationevidence.com/actions/488

Provide artificial nests (wildfowl — artificial/floating islands)

Two studies from North America found that wildfowl used artificial islands and floating rafts and had high nesting success. A study in the UK found that wildfowl preferentially nested on vegetated rather than bare islands. *Assessment: likely to be beneficial (effectiveness 45%; certainty 45%; harms 0%).*

http://www.conservationevidence.com/actions/483

Unknown effectiveness (limited evidence)

Artificially incubate eggs or warm nests

One of two studies found that no kakapo chicks or eggs died of cold when they were artificially warmed when females left the nest. A study from the UK found that great tits were less likely to interrupt their laying sequence if their nest boxes were warmed, but there was no effect on egg or clutch size. *Assessment: unknown effectiveness — limited evidence (effectiveness 26%; certainty 16%; harms 0%).*

http://www.conservationevidence.com/actions/503

Guard nests

We captured four studies describing the effects of guarding nests. One, from Costa Rica, found an increase in scarlet macaw population after nest monitoring and several other interventions. Two studies from Puerto Rico and New Zealand found that nest success was higher, or mortality lower, when nests were monitored. A study from New Zealand found that nest success was high overall when nests were monitored. *Assessment: unknown effectiveness — limited evidence (effectiveness 41%; certainty 24%; harms 0%).*

http://www.conservationevidence.com/actions/506

Provide artificial nests (gamebirds)

A study in China found that approximately 40% of the local population of Cabot's tragopans used nesting platforms. *Assessment: unknown effectiveness — limited evidence (effectiveness 40%; certainty 13%; harms 0%).*

http://www.conservationevidence.com/actions/484

Provide artificial nests (grebes)

A study from the UK found that grebes used nesting rafts in some areas but not others. *Assessment: unknown effectiveness — limited evidence (effectiveness 10%; certainty 9%; harms 0%).*

http://www.conservationevidence.com/actions/479

Provide artificial nests (ibises and flamingos)

A study from Turkey found that ibises moved to a site with artificial breeding ledges. A study from Spain and France found that large numbers of flamingos used artificial nesting islands. *Assessment: unknown effectiveness — limited evidence (effectiveness 42%; certainty 31%; harms 0%).*

http://www.conservationevidence.com/actions/487

Provide artificial nests (parrots)

A study from Costa Rica found that the local population of scarlet macaws increased following the installation of nest boxes along with several other interventions. Five studies from South and Central America and Mauritius found that nest boxes were used by several species of parrots. One study from Peru found that blue-and-yellow macaws only used modified palms, not 'boxes', whilst another study found that scarlet macaws used both PVC and wooden boxes. Four studies from Venezuela and Columbia found that several species rarely, if ever, used nest boxes. Six studies from Central and South America found that parrots nested successfully in nest boxes, with two species showing higher levels of recruitment into the population following nest box erection and another finding that success rates for artificial nests were similar to natural nests. Three studies from South America found that artificial nests had low success rates, in two cases due to poaching. *Assessment: unknown effectiveness — limited evidence (effectiveness 25%; certainty 38%; harms 11%).*

http://www.conservationevidence.com/actions/497

Provide artificial nests (pigeons)

Two studies from the USA and the Netherlands found high use rates and high nesting success of pigeons and doves using artificial nests. *Assessment: unknown effectiveness — limited evidence (effectiveness 30%; certainty 16%; harms 0%).*

http://www.conservationevidence.com/actions/492

Provide artificial nests (rails)

A study from the UK found that common moorhens and common coot readily used artificial nesting islands. *Assessment: unknown effectiveness — limited evidence (effectiveness 30%; certainty 11%; harms 0%).*

http://www.conservationevidence.com/actions/485

Provide artificial nests (rollers)

A study from Spain found that the use of nest boxes by rollers increased over time and varied between habitats. Another study from Spain found no difference in success rates between new and old nest boxes. *Assessment: unknown effectiveness — limited evidence (effectiveness 20%; certainty 20%; harms 0%).*

http://www.conservationevidence.com/actions/494

Provide artificial nests (swifts)

A study from the USA found that Vaux's swifts successfully used nest boxes provided. *Assessment: unknown effectiveness — limited evidence (effectiveness 25%; certainty 16%; harms 0%).*

http://www.conservationevidence.com/actions/495

Provide artificial nests (trogons)

A small study from Guatemala found that at least one resplendent quetzal nested in nest boxes provided. *Assessment: unknown effectiveness — limited evidence (effectiveness 19%; certainty 11%; harms 0%).*

http://www.conservationevidence.com/actions/493

Provide artificial nests (waders)

Two studies from the USA and the UK found that waders used artificial islands and nesting sites. *Assessment: unknown effectiveness — limited evidence (effectiveness 25%; certainty 20%; harms 0%).*

http://www.conservationevidence.com/actions/486

Provide artificial nests (woodpeckers)

Four studies from the USA found local increases in red-cockaded woodpecker populations or the successful colonisation of new areas following the installation of 'cavity inserts'. One study also found that the productivity of birds using the inserts was higher than the regional average. Two studies from the USA found that red-cockaded woodpeckers used cavity inserts, in one case more frequently than making their own holes or using natural cavities. One study from the USA found that woodpeckers roosted, but did not nest, in nest boxes. Five studies from the USA found that some woodpeckers excavated holes in artificial snags but only roosted in excavated holes or nest boxes. *Assessment: unknown effectiveness — limited evidence (effectiveness 35%; certainty 39%; harms 0%).*

http://www.conservationevidence.com/actions/496

Provide nesting habitat for birds that is safe from extreme weather

Two of three studies found that nesting success of waders and terns was no higher on raised areas of nesting substrate, with one finding that similar numbers were lost to flooding. The third study found that Chatham Island oystercatchers used raised nest platforms, but did not report on nesting success. *Assessment: unknown effectiveness — limited evidence (effectiveness 28%; certainty 23%; harms 0%).*

http://www.conservationevidence.com/actions/504

Provide nesting material for wild birds

One of two studies found that wild birds took nesting material provided; the other found only very low rates of use. *Assessment: unknown effectiveness — limited evidence (effectiveness 11%; certainty 9%; harms 0%).*

http://www.conservationevidence.com/actions/501

Remove vegetation to create nesting areas

Two out of six studies found increases in population sizes at seabird and wader colonies after vegetation was cleared and a third found that an entire colony moved to a new site that was cleared of vegetation. Two of these studies found that several interventions were used at once. Two studies found that gulls and terns used plots cleared of vegetation, one of these found that nesting densities were higher on partially-cleared plots than totally cleared, or uncleared, plots. One study found that tern nesting success was higher on plots after they were cleared of vegetation and other interventions were used. *Assessment: unknown effectiveness — limited evidence (effectiveness 45%; certainty 28%; harms 10%).*

http://www.conservationevidence.com/actions/505

Repair/support nests to support breeding

A study from Puerto Rico found that no chicks died from chilling after nine nests were repaired to prevent water getting in. *Assessment: unknown effectiveness — limited evidence (effectiveness 20%; certainty 10%; harms 0%).*

http://www.conservationevidence.com/actions/502

Use differently-coloured artificial nests

A study from the USA found that two bird species (a thrush and a pigeon) both showed colour preferences for artificial nests, but that these preferences differed between species. In each case, clutches in the preferred colour nest were less successful than those in the other colour. *Assessment: unknown effectiveness — limited evidence (effectiveness 3%; certainty 9%; harms 0%).*

http://www.conservationevidence.com/actions/500

3.14.3 Foster chicks in the wild

Based on the collated evidence, what is the current assessment of the effectiveness of interventions for fostering chicks in the wild?	
Likely to be beneficial	• Foster eggs or chicks with wild conspecifics: raptors • Foster eggs or chicks with wild non-conspecifics (cross-fostering): songbirds
Unknown effectiveness (limited evidence)	• Foster eggs or chicks with wild conspecifics: bustards • Foster eggs or chicks with wild conspecifics: cranes • Foster eggs or chicks with wild conspecifics: gannets and boobies • Foster eggs or chicks with wild conspecifics: owls • Foster eggs or chicks with wild conspecifics: parrots • Foster eggs or chicks with wild conspecifics: vultures • Foster eggs or chicks with wild conspecifics: waders • Foster eggs or chicks with wild conspecifics: woodpeckers • Foster eggs or chicks with wild non-conspecifics (cross-fostering): cranes • Foster eggs or chicks with wild non-conspecifics (cross-fostering): ibises • Foster eggs or chicks with wild non-conspecifics (cross-fostering): petrels and shearwaters • Foster eggs or chicks with wild non-conspecifics (cross-fostering): waders

Likely to be beneficial

Foster eggs or chicks with wild conspecifics (raptors)

Ten out of 11 studies from across the world found that fostering raptor chicks to wild conspecifics had high success rates. A single study from the USA found that only one of six eggs fostered to wild eagle nests hatched and was raised. A study from Spain found that Spanish imperial eagle chicks were no more likely to survive to fledging if they were transferred to foster nests from three chick broods (at high risk from siblicide). A study from Spain found that young (15–20 day old) Montagu's harrier chicks were successfully adopted, but three older (27–29 day old) chicks were rejected. *Assessment: likely to be beneficial (effectiveness 70%; certainty 60%; harms 10%).*

http://www.conservationevidence.com/actions/510

Foster eggs or chicks with wild non-conspecifics (cross-fostering) (songbirds)

A study from the USA found that the survival of cross-fostered yellow warbler chicks was lower than previously-published rates for the species. A study from Norway found that the success of cross-fostering small songbirds varied depending on the species of chick and foster birds but recruitment was the same or higher than control chicks. The pairing success of cross-fostered chicks varied depending on species of chick and foster birds. *Assessment: likely to be beneficial (effectiveness 45%; certainty 45%; harms 10%).*

http://www.conservationevidence.com/actions/520

Unknown effectiveness (limited evidence)

Foster eggs or chicks with wild conspecifics (bustards)

A small study in Saudi Arabia found that a captive-bred egg was successfully fostered to a female in the wild. *Assessment: unknown effectiveness — limited evidence (effectiveness 20%; certainty 5%; harms 0%).*

http://www.conservationevidence.com/actions/513

Foster eggs or chicks with wild conspecifics (cranes)

A small study in Canada found high rates of fledging for whooping crane eggs fostered to first time breeders. *Assessment: unknown effectiveness — limited evidence (effectiveness 26%; certainty 11%; harms 0%).*

http://www.conservationevidence.com/actions/512

Foster eggs or chicks with wild conspecifics (gannets and boobies)

A small study in Australia found that gannet chicks were lighter, and hatching and fledging success lower in nests which had an extra egg or chick added. However, overall productivity was non-significantly higher in experimental nests. *Assessment: unknown effectiveness — limited evidence (effectiveness 9%; certainty 11%; harms 0%).*

http://www.conservationevidence.com/actions/507

Foster eggs or chicks with wild conspecifics (owls)

A study in the USA found high fledging rates for barn owl chicks fostered to wild pairs. A study from Canada found that captive-reared burrowing owl chicks fostered to wild nests did not have lower survival or growth rates than wild chicks. *Assessment: unknown effectiveness — limited evidence (effectiveness 35%; certainty 21%; harms 0%).*

http://www.conservationevidence.com/actions/511

Foster eggs or chicks with wild conspecifics (parrots)

A study from Venezuela found that yellow-shouldered Amazon chicks had high fledging rates when fostered to conspecific nests in the wild. A second study from Venezuela found lower poaching rates of yellow-shouldered Amazons when chicks were moved to foster nests closer to a field base. *Assessment: unknown effectiveness — limited evidence (effectiveness 30%; certainty 14%; harms 0%).*

http://www.conservationevidence.com/actions/515

Foster eggs or chicks with wild conspecifics (vultures)

Two small studies in Italy and the USA found that single chicks were successfully adopted by foster conspecifics, although in one case this led to the death of one of the foster parents' chicks. *Assessment: unknown effectiveness — limited evidence (effectiveness 30%; certainty 15%; harms 41%).*

http://www.conservationevidence.com/actions/509

Foster eggs or chicks with wild conspecifics (waders)

Two small trials in North America found that piping plovers accepted chicks introduced into their broods, although in one case the chick died. A study from New Zealand found that survival of fostered black stilts was higher for birds fostered to conspecifics rather than a closely related species. *Assessment: unknown effectiveness — limited evidence (effectiveness 29%; certainty 9%; harms 0%).*

http://www.conservationevidence.com/actions/508

Foster eggs or chicks with wild conspecifics (woodpeckers)

Three studies from the USA found that red-cockaded woodpecker chicks fostered to conspecifics had high fledging rates. One small study found that fostered chicks survived better than chicks translocated with their parents. *Assessment: unknown effectiveness — limited evidence (effectiveness 41%; certainty 29%; harms 0%).*

http://www.conservationevidence.com/actions/514

Foster eggs or chicks with wild non-conspecifics (cross-fostering) (cranes)

Two studies from the USA found low fledging success for cranes fostered to non-conspecifics' nests. *Assessment: unknown effectiveness — limited evidence (effectiveness 14%; certainty 35%; harms 10%).*

http://www.conservationevidence.com/actions/519

Foster eggs or chicks with wild non-conspecifics (cross-fostering) (ibises)

A 2007 literature review describes attempting to foster northern bald ibis chicks with cattle egrets as unsuccessful. *Assessment: unknown effectiveness — limited evidence (effectiveness 0%; certainty 10%; harms 0%).*

http://www.conservationevidence.com/actions/518

Foster eggs or chicks with wild non-conspecifics (cross-fostering) (petrels and shearwaters)

A study from Hawaii found that Newell's shearwater eggs fostered to wedge-tailed shearwater nests had high fledging rates. *Assessment: unknown effectiveness — limited evidence (effectiveness 45%; certainty 6%; harms 0%).*

http://www.conservationevidence.com/actions/516

Foster eggs or chicks with wild non-conspecifics (cross-fostering) (waders)

A study from the USA found that killdeer eggs incubated and raised by spotted sandpipers had similar fledging rates to parent-reared birds. A study from New Zealand found that cross-fostering black stilt chicks to black-winged stilt nests increased nest success, but cross-fostered chicks had lower success than chicks fostered to conspecifics' nests. *Assessment: unknown effectiveness — limited evidence (effectiveness 35%; certainty 30%; harms 0%).*

http://www.conservationevidence.com/actions/517

3.14.4 Provide supplementary food

Based on the collated evidence, what is the current assessment of the effectiveness of interventions for providing supplementary food?	
Beneficial	• Provide supplementary food to increase adult survival: songbirds
Likely to be beneficial	• Place feeders close to windows to reduce collisions • Provide calcium supplements to increase survival or reproductive success • Provide supplementary food to increase adult survival: cranes • Provide supplementary food to increase reproductive success: gulls, terns and skuas • Provide supplementary food to increase reproductive success: owls • Provide supplementary food to increase reproductive success: raptors • Provide supplementary food to increase reproductive success: songbirds
Unknown effectiveness (limited evidence)	• Provide perches to improve foraging success • Provide supplementary food through the establishment of food populations • Provide supplementary food to allow the rescue of a second chick • Provide supplementary food to increase adult survival: gamebirds • Provide supplementary food to increase adult survival: gulls, terns and skuas • Provide supplementary food to increase adult survival: hummingbirds • Provide supplementary food to increase adult survival: nectar-feeding songbirds

	• Provide supplementary food to increase adult survival: pigeons • Provide supplementary food to increase adult survival: raptors • Provide supplementary food to increase adult survival: vultures • Provide supplementary food to increase adult survival: waders • Provide supplementary food to increase adult survival: wildfowl • Provide supplementary food to increase adult survival: woodpeckers • Provide supplementary food to increase reproductive success: auks • Provide supplementary food to increase reproductive success: gamebirds • Provide supplementary food to increase reproductive success: gannets and boobies • Provide supplementary food to increase reproductive success: ibises • Provide supplementary food to increase reproductive success: kingfishers • Provide supplementary food to increase reproductive success: parrots • Provide supplementary food to increase reproductive success: petrels • Provide supplementary food to increase reproductive success: pigeons • Provide supplementary food to increase reproductive success: rails and coots • Provide supplementary food to increase reproductive success: vultures • Provide supplementary food to increase reproductive success: waders • Provide supplementary food to increase reproductive success: wildfowl • Provide supplementary water to increase survival or reproductive success

Beneficial

Provide supplementary food to increase adult survival (songbirds)

Seven studies from Europe and the USA found higher densities or larger populations of songbird species in areas close to supplementary food. Six studies from Europe, Canada and Japan found that population trends or densities were no different between fed and unfed areas. Four studies from around the world found that birds had higher survival when supplied with supplementary food. However, in two studies this was only apparent in some individuals or species and one study from the USA found that birds with feeding stations in their territories had lower survival. Six studies from Europe and the USA found that birds supplied with supplementary food were in better physical condition than unfed birds. However, in four studies this was only true for some individuals, species or seasons. Two studies investigated the effect of feeding on behaviours: one in the USA found that male birds spent more time singing when supplied with food and one in Sweden found no behavioural differences between fed and unfed birds. Thirteen studies from the UK, Canada and the USA investigated use of feeders. Four studies from the USA and the UK found high use of supplementary food, with up to 21% of birds' daily energy needs coming from feeders. However, another UK study found very low use of food. The timing of peak feeder use varied. Two trials from the UK found that the use of feeders increased with distance to houses and decreased with distance to cover. Two studies in Canada and the UK, found that preferences for feeder locations and positions varies between species. *Assessment: beneficial (effectiveness 65%; certainty 75%; harms 0%).*

http://www.conservationevidence.com/actions/552

Likely to be beneficial

Place feeders close to windows to reduce collisions

A randomised, replicated and controlled study in the USA found that fewer birds hit windows, and fewer were killed, when feeders were placed close to windows, compared to when they were placed further away. *Assessment: likely to be beneficial (effectiveness 44%; certainty 43%; harms 0%).*

http://www.conservationevidence.com/actions/557

Provide calcium supplements to increase survival or reproductive success

Eight of 13 studies (including a literature review) from across the world found some positive effects of calcium provisioning on birds' productivites (six studies) or health (two studies). Six studies (including the review) found no evidence for positive effects on some of the species studied. One study from Europe found that birds at polluted sites took more calcium supplement than those at cleaner sites. *Assessment: likely to be beneficial (effectiveness 55%; certainty 50%; harms 0%).*

http://www.conservationevidence.com/actions/559

Provide supplementary food to increase adult survival (cranes)

A study from Japan and a global literature review found that local crane populations increased after the provision of supplementary food. *Assessment: likely to be beneficial (effectiveness 40%; certainty 40%; harms 15%).*

http://www.conservationevidence.com/actions/547

Provide supplementary food to increase reproductive success (gulls, terns and skuas)

Four studies of three experiments from Europe and Alaska found that providing supplementary food increased fledging success or chick survival in two gull species, although a study from the UK found that this was only true for one of two islands. One study from the Antarctic found no effect of feeding parent skuas on productivity. One study from Alaska found increased chick growth when parents were fed but a study from the Antarctic found no such increase. *Assessment: likely to be beneficial (effectiveness 42%; certainty 41%; harms 0%).*

http://www.conservationevidence.com/actions/525

Provide supplementary food to increase reproductive success (owls)

Two replicated, controlled trials from Europe and the USA found that owls supplied with supplementary food had higher hatching and fledging rates. The European study, but not the American, also found that fed pairs laid earlier and had larger clutches. The study in the USA also found that owls were no more likely to colonise nest boxes provided with supplementary food. *Assessment: likely to be beneficial (effectiveness 50%; certainty 42%; harms 0%).*

http://www.conservationevidence.com/actions/533

Provide supplementary food to increase reproductive success (raptors)

A small study in Italy described a small increase in local kite populations following the installation of a feeding station. Four European studies found that kestrels and Eurasian sparrowhawks laid earlier than control birds when supplied with supplementary food. Three studies from the USA and Europe found higher chick survival or condition when parents were supplied with food, whilst three from Europe found fed birds laid larger clutches and another found that fed male hen harriers bred with more females than control birds. Four studies from across the world found no evidence that feeding increased breeding frequency, clutch size, laying date, eggs size or hatching or fledging success. A study from Mauritius found uncertain effects of feeding on Mauritius kestrel reproduction. There was some evidence that the impact of feeding was lower in years with peak numbers of prey species. *Assessment: likely to be beneficial (effectiveness 55%; certainty 52%; harms 0%).*

http://www.conservationevidence.com/actions/532

Provide supplementary food to increase reproductive success (songbirds)

Two studies from the USA found evidence for higher population densities of magpies and American blackbirds in areas provided with supplementary food, whilst two studies from the UK and Canada found that population densities were not affected by feeding. Twelve studies from across the world found that productivity was higher for fed birds than controls. Eleven studies from Europe and the USA found that fed birds had the same, or even lower, productivity or chick survival than control birds. Nine studies from Europe and North America found that the eggs of fed birds were larger or heavier, or that the chicks of fed birds were in better physical condition. However, eight studies from across the world found no evidence for better condition or increased size in the eggs or chicks of fed birds. Six studies from across the world found that food-supplemented pairs laid larger clutches, whilst 14 studies from Europe and North America found that fed birds did not lay larger clutches. Fifteen studies from across the world found that birds supplied with supplementary food began nesting earlier than controls, although in two cases only certain individuals, or those in particular habitats, laid earlier. One study found that fed birds had shorter incubations than controls whilst another found that fed birds re-nested quicker and had shorter second incubations. Four studies from the USA and Europe found that fed birds did not lay any earlier than controls. Seven studies from across the world found that fed parent birds showed positive behavioural responses to feeding. However, three studies from across the world found neutral or negative responses to feeding. *Assessment: likely to be beneficial (effectiveness 51%; certainty 85%; harms 6%).*

http://www.conservationevidence.com/actions/537

Unknown effectiveness (limited evidence)

Provide perches to improve foraging success

One of four studies, from Sweden, found that raptors used clearcuts provided with perches more than clearcuts without perches. Two studies found that birds used perches provided, but a controlled study from the USA found that shrikes did not alter foraging behaviour when perches were present. *Assessment: unknown effectiveness — limited evidence (effectiveness 45%; certainty 30%; harms 0%).*

http://www.conservationevidence.com/actions/556

Provide supplementary food through the establishment of food populations

One of four studies that established prey populations found that wildfowl fed on specially-planted rye grass. Two studies found that cranes in the USA and owls in Canada did not respond to established prey populations. A study from Sweden found that attempts to increase macroinvertebrate numbers for wildfowl did not succeed. *Assessment: unknown effectiveness — limited evidence (effectiveness 9%; certainty 26%; harms 0%)..*

http://www.conservationevidence.com/actions/555

Provide supplementary food to allow the rescue of a second chick

A study from Spain found that second chicks from lammergeier nests survived longer if nests were provided with food, in one case allowing a chick to be rescued. *Assessment: unknown effectiveness — limited evidence (effectiveness 15%; certainty 14%; harms 0%).*

http://www.conservationevidence.com/actions/541

Provide supplementary food to increase adult survival (gamebirds)

Two European studies found increased numbers of birds in fed areas, compared to unfed areas. There was only an increase in the overall population in the study area in one of these studies. Of four studies in the USA on northern bobwhites, one found that birds had higher overwinter survival in fed areas, one found lower survival, one found fed birds had higher body fat percentages and a literature review found no overall effect of feeding. *Assessment: unknown effectiveness — limited evidence (effectiveness 49%; certainty 38%; harms 0%).*

http://www.conservationevidence.com/actions/544

Provide supplementary food to increase adult survival (gulls, terns and skuas)

A study in the Antarctic found that fed female south polar skuas lost more weight whilst feeding two chicks than unfed birds. There was no difference for birds with single chicks, or male birds. *Assessment: unknown effectiveness — limited evidence (effectiveness 0%; certainty 20%; harms 10%).*

http://www.conservationevidence.com/actions/548

Provide supplementary food to increase adult survival (hummingbirds)

Four studies from the USA found that three species of hummingbird preferred higher concentrations of sucrose, consuming more and visiting feeders more frequently. A study from the USA found that hummingbirds preferentially fed on sugar solutions over artificial sweeteners, and that the viscosity of these solutions did not affect their consumption. Two studies from Mexico and Argentina found that four species showed preferences for sucrose over fructose or glucose and sucrose over a sucrose-glucose mix, but no preference for sucrose over a glucose-fructose mix. A study from the USA found that birds showed a preference for red-dyed sugar solutions over five other colours. A study from the USA found that rufous hummingbirds preferentially fed on feeders that were placed higher. *Assessment: unknown effectiveness — limited evidence (effectiveness 10%; certainty 24%; harms 0%).*

http://www.conservationevidence.com/actions/550

Provide supplementary food to increase adult survival (nectar-feeding songbirds)

Two studies from Australia and New Zealand found that ten species of honeyeaters and stitchbirds readily used feeders supplying sugar solutions, with seasonal variations varying between species. A series of *ex situ* trials using southern African birds found that most species preferred sucrose solutions over glucose or fructose. One study found that sunbirds and sugarbirds only showed such a preference at low concentrations. Two studies found that two species showed preferences for sucrose when comparing 20% solutions, although a third species did not show this preference. All species rejected solutions with xylose added. A final study found that sucrose preferences were only apparent at equicalorific concentrations high enough for birds to subsist on. *Assessment: unknown effectiveness — limited evidence (effectiveness 10%; certainty 23%; harms 0%).*

http://www.conservationevidence.com/actions/553

Provide supplementary food to increase adult survival (pigeons)

The first of two studies of a recently-released pink pigeon population on Mauritius found that fewer than half the birds took supplementary food. However, the later study found that almost all birds used supplementary feeders. *Assessment: unknown effectiveness — limited evidence (effectiveness 10%; certainty 19%; harms 0%).*

http://www.conservationevidence.com/actions/549

Provide supplementary food to increase adult survival (raptors)

Two studies in the USA found that nesting northern goshawks were significantly heavier in territories supplied with supplementary food, compared with those from unfed territories. *Assessment: unknown effectiveness— limited evidence (effectiveness 30%; certainty 23%; harms 0%).*

http://www.conservationevidence.com/actions/546

Provide supplementary food to increase adult survival (vultures)

A study from Spain found a large increase in griffon vulture population in the study area following multiple interventions including supplementary feeding. Two studies from the USA and Israel found that vultures fed on the carcasses provided for them. In the study in Israel vultures were sometimes dominated by larger species at a feeding station supplied twice a month, but not at one supplied every day. *Assessment: unknown effectiveness — limited evidence (effectiveness 18%; certainty 18%; harms 0%).*

http://www.conservationevidence.com/actions/545

Provide supplementary food to increase adult survival (waders)

A study in Northern Ireland found that waders fed on millet seed when provided, but were dominated by other ducks when larger seeds were provided. *Assessment: unknown effectiveness — limited evidence (effectiveness 22%; certainty 9%; harms 0%).*

http://www.conservationevidence.com/actions/543

Provide supplementary food to increase adult survival (wildfowl)

Two studies from Canada and Northern Ireland found that five species of wildfowl readily consumed supplementary grains and seeds. The Canadian study found that fed birds were heavier and had larger hearts or flight muscles or more body fat than controls. *Assessment: unknown effectiveness — limited evidence (effectiveness 14%; certainty 15%; harms 0%).*

http://www.conservationevidence.com/actions/542

Provide supplementary food to increase adult survival (woodpeckers)

One replicated, controlled study from the USA found that 12 downy woodpeckers supplied with supplementary food had higher nutritional statuses than unfed birds. However, two analyses of a replicated, controlled study of 378 downy woodpeckers from the USA found that they did not have higher survival rates or nutritional statuses than unfed birds. *Assessment: unknown effectiveness — limited evidence (effectiveness 10%; certainty 30%; harms 0%).*

http://www.conservationevidence.com/actions/551

Provide supplementary food to increase reproductive success (auks)

Two replicated studies from the UK found that Atlantic puffin chicks provided with supplementary food were significantly heavier than control chicks, but fed chicks fledged at the same time as controls. A randomised, replicated and controlled study from Canada found that tufted puffin chicks supplied with supplementary food fledged later than controls and that fed chicks had faster growth by some, but not all, metrics. *Assessment: unknown effectiveness — limited evidence (effectiveness 30%; certainty 38%; harms 0%).*

http://www.conservationevidence.com/actions/524

Provide supplementary food to increase reproductive success (gamebirds)

A controlled study in Tibet found that Tibetan eared pheasants fed supplementary food laid significantly larger eggs and clutches than control birds. Nesting success and laying dates were not affected. *Assessment: unknown effectiveness — limited evidence (effectiveness 23%; certainty 10%; harms 0%).*

http://www.conservationevidence.com/actions/527

Provide supplementary food to increase reproductive success (gannets and boobies)

A small controlled study in Australia found that Australasian gannet chicks were significantly heavier if they were supplied with supplementary food, but only in one of two years. Fledging success of fed nests was also higher, but not significantly so. A randomised replicated and controlled study in the Galapagos Islands found that fed female Nazca boobies were more likely to produce two-egg clutches, and that second eggs were significantly heavier. *Assessment: unknown effectiveness — limited evidence (effectiveness 33%; certainty 25%; harms 0%).*

http://www.conservationevidence.com/actions/523

Provide supplementary food to increase reproductive success (ibises)

A study from China found that breeding success of crested ibis was correlated with the amount of supplementary food provided, although no comparison was made with unfed nests. *Assessment: unknown effectiveness — limited evidence (effectiveness 25%; certainty 11%; harms 0%).*

http://www.conservationevidence.com/actions/530

Provide supplementary food to increase reproductive success (kingfishers)

A controlled study in the USA found that belted kingfishers supplied with food had heavier nestlings and were more likely to renest. There was mixed evidence for the effect of feeding on laying date. *Assessment: unknown effectiveness — limited evidence (effectiveness 33%; certainty 13%; harms 0%).*

http://www.conservationevidence.com/actions/534

Provide supplementary food to increase reproductive success (parrots)

Two studies from New Zealand found evidence that providing supplementary food for kakapos increased the number of breeding attempts made, whilst a third study found that birds provided with specially-formulated pellets appeared to have larger clutches than those fed on nuts. One study found no evidence that providing food increased the number of nesting attempts. *Assessment: unknown effectiveness — limited evidence (effectiveness 33%; certainty 11%; harms 0%).*

http://www.conservationevidence.com/actions/536

Provide supplementary food to increase reproductive success (petrels)

A replicated controlled study in Australia found that Gould's petrel chicks provided with supplementary food had similar fledging rates to both control and hand-reared birds, but were significantly heavier than other birds. *Assessment: unknown effectiveness — limited evidence (effectiveness 19%; certainty 14%; harms 0%).*

http://www.conservationevidence.com/actions/522

Provide supplementary food to increase reproductive success (pigeons)

A study in the UK found no differences in reproductive parameters of European turtle doves between years when food was supplied and those when it was not. *Assessment: unknown effectiveness — limited evidence (effectiveness 0%; certainty 21%; harms 0%).*

http://www.conservationevidence.com/actions/535

Provide supplementary food to increase reproductive success (rails and coots)

A small trial in the USA found that fed American coots laid heavier eggs, but not larger clutches, than controls. However, a randomised, replicated and controlled study in Canada found that clutch size, but not egg size, was larger in fed American coot territories. The Canadian study also found that coots laid earlier when fed, whilst a replicated trial from the UK found there was a shorter interval between common moorhens clutches in fed territories, but that fed birds were no more likely to produce second broods. *Assessment: unknown effectiveness — limited evidence (effectiveness 33%; certainty 26%; harms 0%).*

http://www.conservationevidence.com/actions/528

Provide supplementary food to increase reproductive success (vultures)

Two studies from the USA and Greece found that there were local increases in two vulture populations following the provision of food in the area. A study from Israel found that a small, regularly supplied feeding station could provide sufficient food for breeding Egyptian vultures. A study from Italy found that a small population of Egyptian vultures declined following the provision of food, and only a single vulture was seen at the feeding station. *Assessment: unknown effectiveness — limited evidence (effectiveness 30%; certainty 24%; harms 0%).*

http://www.conservationevidence.com/actions/531

Provide supplementary food to increase reproductive success (waders)

A small controlled trial from the Netherlands found that Eurasian oystercatchers did not produce larger replacement eggs if provided with supplementary food. Instead their eggs were smaller than the first clutch, whereas control females laid larger replacement eggs. *Assessment: unknown effectiveness — limited evidence (effectiveness 0%; certainty 10%; harms 0%).*

http://www.conservationevidence.com/actions/529

Provide supplementary food to increase reproductive success (wildfowl)

A small randomised controlled *ex situ* study from Canada found faster growth and higher weights for fed greater snow goose chicks than unfed ones, but no differences in mortality rates. *Assessment: unknown effectiveness — limited evidence (effectiveness 30%; certainty 10%; harms 0%).*

http://www.conservationevidence.com/actions/526

Provide supplementary water to increase survival or reproductive success

A controlled study from Morocco found that northern bald ibises provided with supplementary water had higher reproductive success than those a long way from water sources. *Assessment: unknown effectiveness — limited evidence (effectiveness 43%; certainty 14%; harms 0%).*

http://www.conservationevidence.com/actions/558

3.14.5 Translocations

Based on the collated evidence, what is the current assessment of the effectiveness of interventions for translocations?	
Beneficial	• Translocate birds to re-establish populations or increase genetic variation (birds in general) • Translocate birds to re-establish populations or increase genetic variation: raptors

Likely to be beneficial	• Translocate birds to re-establish populations or increase genetic variation: parrots • Translocate birds to re-establish populations or increase genetic variation: pelicans • Translocate birds to re-establish populations or increase genetic variation: petrels and shearwaters • Translocate birds to re-establish populations or increase genetic variation: rails • Translocate birds to re-establish populations or increase genetic variation: songbirds • Translocate birds to re-establish populations or increase genetic variation: wildfowl • Translocate birds to re-establish populations or increase genetic variation: woodpeckers • Use decoys to attract birds to new sites • Use techniques to increase the survival of species after capture • Use vocalisations to attract birds to new sites
Trade-off between benefit and harms	• Translocate birds to re-establish populations or increase genetic variation: gamebirds
Unknown effectiveness (limited evidence)	• Alter habitats to encourage birds to leave • Ensure translocated birds are familiar with each other before release • Translocate birds to re-establish populations or increase genetic variation: auks • Translocate birds to re-establish populations or increase genetic variation: herons, storks and ibises • Translocate birds to re-establish populations or increase genetic variation: megapodes • Translocate birds to re-establish populations or increase genetic variation: owls • Translocate nests to avoid disturbance
No evidence found (no assessment)	• Ensure genetic variation to increase translocation success

Beneficial

Translocate birds to re-establish populations or increase genetic variation (birds in general)

A review of 239 bird translocation programmes found 63–67% resulted in establishment of a self-sustaining population. *Assessment: beneficial (effectiveness 64%; certainty 65%; harms 0%).*

http://www.conservationevidence.com/actions/566

Translocate birds to re-establish populations or increase genetic variation (raptors)

Six studies of three translocation programmes in the UK and the USA found that all successfully established populations of white-tailed eagles, red kites and ospreys. A study in Spain found high survival of translocated Montagu's harrier fledglings. *Assessment: beneficial (effectiveness 65%; certainty 66%; harms 0%).*

http://www.conservationevidence.com/actions/574

Likely to be beneficial

Translocate birds to re-establish populations or increase genetic variation (parrots)

Three studies of two translocation programmes from the Pacific and New Zealand found that populations of parrots successfully established on islands after translocation. Survival of translocated birds ranged from 41% to 98% globally. Despite high survival, translocated kakapos in New Zealand had very low reproductive output. *Assessment: likely to be beneficial (effectiveness 50%; certainty 60%; harms 10%).*

http://www.conservationevidence.com/actions/578

Translocate birds to re-establish populations or increase genetic variation (pelicans)

Two reviews of a pelican translocation programme in the USA found high survival of translocated nestlings and rapid target population growth.

Some growth may have been due to additional immigration from the source populations. *Assessment: likely to be beneficial (effectiveness 55%; certainty 49%; harms 0%).*

http://www.conservationevidence.com/actions/569

Translocate birds to re-establish populations or increase genetic variation (petrels and shearwaters)

Three studies from Australia and New Zealand found that colonies of burrow-nesting petrels and shearwaters were successfully established following the translocation and hand-rearing of chicks. *Assessment: likely to be beneficial (effectiveness 60%; certainty 50%; harms 0%).*

http://www.conservationevidence.com/actions/568

Translocate birds to re-establish populations or increase genetic variation (rails)

Three studies of two translocation programmes in the Seychelles and New Zealand found high survival rates among translocated rail. All three studies round that the birds bred successfully. *Assessment: likely to be beneficial (effectiveness 54%; certainty 44%; harms 14%).*

http://www.conservationevidence.com/actions/573

Translocate birds to re-establish populations or increase genetic variation (songbirds)

Nine studies from across the world, including a review of 31 translocation attempts, found that translocations led to the establishment of songbird populations. Eight studies were on islands. Three studies reported on translocations that failed to establish populations. One study found nesting success decreased as the latitudinal difference between source area and release site increased. *Assessment: likely to be beneficial (effectiveness 50%; certainty 68%; harms 0%).*

http://www.conservationevidence.com/actions/580

Translocate birds to re-establish populations or increase genetic variation (wildfowl)

Three studies of two duck translocation programmes in New Zealand and Hawaii found high survival, breeding and successful establishment of new populations. However a study in the USA found that no ducks stayed at the release site and there was high mortality after release. A study in the USA found wing-clipping prevented female ducks from abandoning their ducklings. *Assessment: likely to be beneficial (effectiveness 42%; certainty 50%; harms 19%).*

http://www.conservationevidence.com/actions/571

Translocate birds to re-establish populations or increase genetic variation (woodpeckers)

Six studies of four programmes found that >50% translocated birds remained at their new sites, and two studies reported large population increases. Birds from four programmes were reported as forming pairs or breeding and one study round translocated nestlings fledged at similar rates to native chicks. All studies were of red-cockaded woodpeckers. *Assessment: likely to be beneficial (effectiveness 51%; certainty 42%; harms 0%).*

http://www.conservationevidence.com/actions/577

Use decoys to attract birds to new sites

Ten studies found that birds nested in areas where decoys were placed or that more birds landed in areas with decoys than control areas. Six studies used multiple interventions at once. One study found that three-dimensional models appeared more effective than two-dimensional ones, and that plastic models were more effective than rag decoys. *Assessment: likely to be beneficial (effectiveness 51%; certainty 45%; harms 0%).*

http://www.conservationevidence.com/actions/586

Use techniques to increase the survival of species after capture

A study from the USA found that providing dark, quiet environments with readily-available food and water increased the survival of small songbirds after capture and the probability that they would adapt to captivity. A study from the USA found that keeping birds warm during transit increased survival. *Assessment: likely to be beneficial (effectiveness 49%; certainty 41%; harms 0%).*

http://www.conservationevidence.com/actions/581

Use vocalisations to attract birds to new sites

Seven out of ten studies from around the world found that seabirds were more likely to nest or land to areas where vocalisations were played, or moved to new nesting areas after vocalisations were played. Four of these studied multiple interventions at once. Three studies found that birds were no more likely to nest or land in areas where vocalisations were played. *Assessment: likely to be beneficial (effectiveness 45%; certainty 50%; harms 0%).*

http://www.conservationevidence.com/actions/585

Trade-off between benefit and harms

Translocate birds to re-establish populations or increase genetic variation (gamebirds)

Three studies from the USA found that translocation of gamebirds led to population establishment or growth or an increase in lekking sites. Four studies from the USA found that translocated birds had high survival, but two found high mortality in translocated birds. Four studies from the USA found breeding rates among translocated birds were high or similar to resident birds. *Assessment: trade-offs between benefits and harms (effectiveness 50%; certainty 47%; harms 35%).*

http://www.conservationevidence.com/actions/572

Unknown effectiveness (limited evidence)

Alter habitats to encourage birds to leave

A study from Canada found that an entire Caspian tern population moved after habitat was altered at the old colony site, alongside several other interventions. *Assessment: unknown effectiveness — limited evidence (effectiveness 20%; certainty 10%; harms 0%).*

http://www.conservationevidence.com/actions/587

Ensure translocated birds are familiar with each other before release

Two studies from New Zealand found no evidence that ensuring birds were familiar with each other increased translocation success. *Assessment:*

unknown effectiveness — limited evidence (effectiveness 0%; certainty 33%; harms 0%).

http://www.conservationevidence.com/actions/582

Translocate birds to re-establish populations or increase genetic variation (auks)

A study in the USA and Canada found that 20% of translocated Atlantic puffins remained in or near the release site, with up to 7% breeding. *Assessment: unknown effectiveness — limited evidence (effectiveness 36%; certainty 38%; harms 0%).*

http://www.conservationevidence.com/actions/570

Translocate birds to re-establish populations or increase genetic variation (herons, storks and ibises)

A study in the USA found that a colony of black-crowned night herons was successfully translocated and bred the year after translocation. *Assessment: unknown effectiveness — limited evidence (effectiveness 44%; certainty 3%; harms 0%).*

http://www.conservationevidence.com/actions/575

Translocate birds to re-establish populations or increase genetic variation (megapodes)

A study from Indonesia found that up to 78% maleo eggs hatched after translocation. *Assessment: unknown effectiveness — limited evidence (effectiveness 49%; certainty 29%; harms 0%).*

http://www.conservationevidence.com/actions/567

Translocate birds to re-establish populations or increase genetic variation (owls)

A small study from New Zealand found that translocating two male boobooks allowed the establishment of a population when they interbred with a Norfolk Island boobook. A study in the USA found high survival amongst burrowing owls translocated as juveniles, although birds were not seen after release. *Assessment: unknown effectiveness — limited evidence (effectiveness 20%; certainty 19%; harms 0%).*

http://www.conservationevidence.com/actions/576

Translocate nests to avoid disturbance

All five studies captured found some success in relocating nests while they were in use, but one found that fewer than half of the burrowing owls studied were moved successfully; a study found that repeated disturbance caused American kestrels to abandon their nest and a study found that one barn swallow abandoned its nest after it was moved. *Assessment: unknown effectiveness — limited evidence (effectiveness 24%; certainty 39%; harms 30%).*

http://www.conservationevidence.com/actions/584

No evidence found (no assessment)

We have captured no evidence for the following interventions:

- Ensure genetic variation to increase translocation success

3.15 Captive breeding, rearing and releases (*ex situ* conservation)

3.15.1 Captive breeding

Based on the collated evidence, what is the current assessment of the effectiveness of interventions for captive breeding?	
Likely to be beneficial	• Artificially incubate and hand-rear birds in captivity: raptors • Artificially incubate and hand-rear birds in captivity: seabirds • Artificially incubate and hand-rear birds in captivity: songbirds • Artificially incubate and hand-rear birds in captivity: waders • Use captive breeding to increase or maintain populations: raptors
Unknown effectiveness (limited evidence)	• Artificially incubate and hand-rear birds in captivity: bustards • Artificially incubate and hand-rear birds in captivity: cranes • Artificially incubate and hand-rear birds in captivity: gamebirds • Artificially incubate and hand-rear birds in captivity: parrots

	• Artificially incubate and hand-rear birds in captivity: penguins • Artificially incubate and hand-rear birds in captivity: rails • Artificially incubate and hand-rear birds in captivity: storks and ibises • Artificially incubate and hand-rear birds in captivity: vultures • Artificially incubate and hand-rear birds in captivity: wildfowl • Freeze semen for artificial insemination • Use artificial insemination in captive breeding • Use captive breeding to increase or maintain populations: bustards • Use captive breeding to increase or maintain populations: cranes • Use captive breeding to increase or maintain populations: pigeons • Use captive breeding to increase or maintain populations: rails • Use captive breeding to increase or maintain populations: seabirds • Use captive breeding to increase or maintain populations: songbirds • Use captive breeding to increase or maintain populations: storks and ibises • Use captive breeding to increase or maintain populations: tinamous • Use puppets to increase the success of hand-rearing • Wash contaminated semen and use it for artificial insemination
Evidence not assessed	• Can captive breeding have deleterious effects on individual fitness?

Likely to be beneficial

Artificially incubate and hand-rear birds in captivity (raptors)

Six studies from across the world found high success rates for artificial incubation and hand-rearing of raptors. A replicated and controlled study from France found that artificially incubated raptor eggs had lower hatching success than parent-incubated eggs but fledging success for hand-reared chicks was similar to wild chicks. A study from Canada found that hand-reared chicks had slower growth and attained a lower weight than parent-reared birds. A replicated study from Mauritius found that hand-rearing of wild eggs had higher success than hand-rearing captive-bred chicks. Three studies that provided methodological comparisons reported that incubation temperature affected hatching success and adding saline to the diet of falcon chicks increased their weight gain. *Assessment: likely to be beneficial (effectiveness 60%; certainty 52%; harms 5%).*

http://www.conservationevidence.com/actions/614

Artificially incubate and hand-rear birds in captivity (seabirds)

Five studies from across the world found evidence for the success of hand-rearing seabirds. One small study in Spain found that one of five hand-reared Audouin's gulls successfully bred in the wild. Four studies found that various petrel species successfully fledged after hand-rearing. One controlled study found that fledging rates of hand-reared birds was similar to parent-reared birds, although a study on a single bird found that the chick fledged at a lower weight and later than parent-reared chicks. *Assessment: likely to be beneficial (effectiveness 67%; certainty 45%; harms 2%).*

http://www.conservationevidence.com/actions/604

Artificially incubate and hand-rear birds in captivity (songbirds)

Four studies from the USA found high rates of success for artificial incubation and hand-rearing of songbirds. One study found that crow chicks fed more food had higher growth rates, but these rates never matched those of wild birds. *Assessment: likely to be beneficial (effectiveness 51%; certainty 44%; harms 1%).*

http://www.conservationevidence.com/actions/616

Artificially incubate and hand-rear birds in captivity (waders)

Three out of four replicated and controlled studies from the USA and New Zealand found that artificially incubated and/or hand-reared waders had higher hatching and fledging success than controls. One study from New Zealand found that hatching success of black stilt was lower for artificially-incubated eggs. *Assessment: likely to be beneficial (effectiveness 64%; certainty 41%; harms 4%).*

http://www.conservationevidence.com/actions/611

Use captive breeding to increase or maintain populations (raptors)

Three small studies and a review from around the world found that raptors bred successfully in captivity. Two of these studies found that wild-caught birds bred in captivity after a few years, with one pair of brown goshawks producing 15 young over four years, whilst a study on bald eagle captive breeding found low fertility in captive-bred eggs, but that birds still produced chicks after a year. A review of Mauritius kestrel captive breeding found that 139 independent young were raised over 12 years from 30 eggs and chicks taken from the wild. An update of the same programme found that hand-reared Mauritius kestrels were less successful if they came from captive-bred eggs compared to wild 'harvested' eggs. *Assessment: likely to be beneficial (effectiveness 50%; certainty 41%; harms 10%).*

http://www.conservationevidence.com/actions/596

Unknown effectiveness (limited evidence)

Artificially incubate and hand-rear birds in captivity (bustards)

Two reviews of a houbara bustard captive breeding programme in Saudi Arabia found no difference in survival between artificially and parentally incubated eggs, and that removing eggs from clutches as they were laid increased the number laid by females. *Assessment: unknown effectiveness — limited evidence (effectiveness 31%; certainty 10%; harms 0%).*

http://www.conservationevidence.com/actions/610

Artificially incubate and hand-rear birds in captivity (cranes)

Two studies from the USA found that hand-reared birds showed normal reproductive behaviour and higher survival than parent-reared birds. *Assessment: unknown effectiveness — limited evidence (effectiveness 76%; certainty 31%; harms 0%).*

http://www.conservationevidence.com/actions/609

Artificially incubate and hand-rear birds in captivity (gamebirds)

A study in Finland found that hand-reared grey partridges did not take off to fly as effectively as wild-caught birds, potentially making them more vulnerable to predation. *Assessment: unknown effectiveness — limited evidence (effectiveness 11%; certainty 10%; harms 50%).*

http://www.conservationevidence.com/actions/607

Artificially incubate and hand-rear birds in captivity (parrots)

Two studies from South America describe the successful hand-rearing of parrot chicks. A review of the kakapo management programme found that chicks could be successfully raised and released, but that eggs incubated from a young age had low success. A study from the USA found that all hand-reared thick-billed parrots died within a month of release: significantly lower survival than for wild-caught birds translocated to the release site. *Assessment: unknown effectiveness — limited evidence (effectiveness 19%; certainty 30%; harms 11%).*

http://www.conservationevidence.com/actions/615

Artificially incubate and hand-rear birds in captivity (penguins)

Two replicated and controlled studies from South Africa found that hand-reared and released African penguins had similar survival and breeding success as birds which were not hand-reared. *Assessment: unknown effectiveness — limited evidence (effectiveness 41%; certainty 15%; harms 0%).*

http://www.conservationevidence.com/actions/605

Artificially incubate and hand-rear birds in captivity (rails)

A controlled study from New Zealand found that post-release survival of hand-reared takahe was as high as wild-reared birds and that six of ten released females raised chicks. *Assessment: unknown effectiveness — limited evidence (effectiveness 64%; certainty 13%; harms 0%).*

http://www.conservationevidence.com/actions/608

Artificially incubate and hand-rear birds in captivity (storks and ibises)

A small study in the USA describes the successful artificial incubation and hand-rearing of two Abdim's stork chicks, whilst a review of northern bald ibis conservation found that only very intensive rearing of a small number of chicks appeared to allow strong bonds, thought to be important for the successful release of birds into the wild, to form between chicks. *Assessment: unknown effectiveness — limited evidence (effectiveness 18%; certainty 10%; harms 0%).*

http://www.conservationevidence.com/actions/612

Artificially incubate and hand-rear birds in captivity (vultures)

A study in Peru found that hand-reared Andean condors had similar survival to parent-reared birds after release into the wild. *Assessment: unknown effectiveness — limited evidence (effectiveness 30%; certainty 10%; harms 0%).*

http://www.conservationevidence.com/actions/613

Artificially incubate and hand-rear birds in captivity (wildfowl)

Two studies in Canada and India found high success rates for hand-rearing buffleheads and bar-headed geese in captivity. Eggs were artificially incubated or incubated under foster parents. A replicated, controlled study in England found that Hawaiian geese (nene) chicks showed less well-adapted behaviours if they were raised without parental contact. *Assessment: unknown effectiveness — limited evidence (effectiveness 50%; certainty 20%; harms 10%).*

http://www.conservationevidence.com/actions/606

Freeze semen for artificial insemination

Two small trials from the USA found that using thawed frozen semen for artificial insemination resulted in low fertility rates. A small trial from the USA found that a cryprotectant increased fertility rates achieved using frozen semen. *Assessment: unknown effectiveness — limited evidence (effectiveness 10%; certainty 10%; harms 45%).*

http://www.conservationevidence.com/actions/602

Use artificial insemination in captive breeding

A replicated study from Saudi Arabia found that artificial insemination could increase fertility in houbara bustards. A study of the same programme and a review found that repeated inseminations increased fertility, with the review arguing that artificial insemination had the potential to be a useful technique. Two studies from the USA found that artificially-inseminated raptors had either zero fertility, or approximately 50%. *Assessment: unknown effectiveness — limited evidence (effectiveness 33%; certainty 21%; harms 0%).*

http://www.conservationevidence.com/actions/601

Use captive breeding to increase or maintain populations (bustards)

Four studies of a captive breeding programme in Saudi Arabia reported that the houbara bustard chicks were successfully raised in captivity, with 285 chicks hatched in the 7th year of the project after 232 birds were used to start the captive population. Captive birds bred earlier and appeared to lay more eggs than wild birds. Forty-six percent of captive eggs hatched and 43% of chicks survived to ten years old. *Assessment: unknown effectiveness — limited evidence (effectiveness 41%; certainty 16%; harms 5%).*

http://www.conservationevidence.com/actions/592

Use captive breeding to increase or maintain populations (cranes)

A study from Canada over 32 years found that whooping cranes successfully bred in captivity eight years after the first eggs were removed from the wild. *Assessment: unknown effectiveness — limited evidence (effectiveness 51%; certainty 17%; harms 6%).*

http://www.conservationevidence.com/actions/591

Use captive breeding to increase or maintain populations (pigeons)

A review of a captive-breeding programme on Mauritius and in the UK found that 42 pink pigeons were successfully bred in captivity. *Assessment: unknown effectiveness — limited evidence (effectiveness 69%; certainty 21%; harms 0%).*

http://www.conservationevidence.com/actions/597

Use captive breeding to increase or maintain populations (rails)

A study from Australia found that three pairs of Lord Howe Island woodhens successfully bred in captivity, with 66 chicks being produced over four years. *Assessment: unknown effectiveness — limited evidence (effectiveness 26%; certainty 11%; harms 5%).*

http://www.conservationevidence.com/actions/590

Use captive breeding to increase or maintain populations (seabirds)

A study from Spain found that a single pair of Audouin's gulls successfully bred in captivity. *Assessment: unknown effectiveness — limited evidence (effectiveness 20%; certainty 4%; harms 5%).*

http://www.conservationevidence.com/actions/589

Use captive breeding to increase or maintain populations (songbirds)

Three studies from Australia and the USA found that three species of songbird bred successfully in captivity. Four out of five pairs of wild-bred, hand-reared puaiohi formed pairs and laid a total of 39 eggs and a breeding population of helmeted honeyeaters was successfully established through a breeding programme. Only one pair of loggerhead shrikes formed pairs from eight wild birds caught and their first clutch died. *Assessment: unknown effectiveness — limited evidence (effectiveness 77%; certainty 31%; harms 5%).*

http://www.conservationevidence.com/actions/598

Use captive breeding to increase or maintain populations (storks and ibises)

We captured a small study and a review both from the USA describing the captive breeding of storks. The study found that a pair bred; the review found that only seven of 19 species had been successfully bred in captivity. A review of bald ibis conservation found that 1,150 birds had been produced in captivity from 150 founders over 20 years. However, some projects had failed, and a study from Turkey found that captive birds had lower productivity than wild birds. *Assessment: unknown effectiveness — limited evidence (effectiveness 31%; certainty 30%; harms 8%).*

http://www.conservationevidence.com/actions/595

Use captive breeding to increase or maintain populations (tinamous)

A replicated study from Costa Rica found that great tinamous successfully bred in captivity, with similar reproductive success to wild birds. *Assessment: unknown effectiveness — limited evidence (effectiveness 51%; certainty 15%; harms 5%).*

http://www.conservationevidence.com/actions/588

Use puppets to increase the success of hand-rearing

Three studies from the USA and Saudi Arabia found that crows and bustards raised using puppets did not have higher survival, dispersal or growth than chicks hand-reared conventionally. *Assessment: unknown effectiveness — limited evidence (effectiveness 4%; certainty 20%; harms 0%).*

http://www.conservationevidence.com/actions/617

Wash contaminated semen and use it for artificial insemination

A replicated, controlled study from Spain found that washed, contaminated semen could be used to successfully inseminate raptors. *Assessment: unknown effectiveness — limited evidence (effectiveness 31%; certainty 15%; harms 0%).*

http://www.conservationevidence.com/actions/603

Evidence not assessed

Can captive breeding have deleterious effects?

We captured no studies investigating the effects of captive-breeding on fitness. Three studies using wild and captive populations or museum specimens found physiological or genetic changes in populations that had been bred in captivity. One found that changes were more likely to be caused by extremely low population levels than by captivity.

http://www.conservationevidence.com/actions/599

3.15.2 Release captive-bred individuals

Based on the collated evidence, what is the current assessment of the effectiveness of interventions for captive breeding?	
Likely to be beneficial	• Provide supplementary food after release • Release captive-bred individuals into the wild to restore or augment wild populations: cranes • Release captive-bred individuals into the wild to restore or augment wild populations: raptors • Release captive-bred individuals into the wild to restore or augment wild populations: songbirds • Release captive-bred individuals into the wild to restore or augment wild populations: vultures
Unknown effectiveness (limited evidence)	• Clip birds' wings on release • Release birds as adults or sub-adults not juveniles • Release birds in groups • Release captive-bred individuals into the wild to restore or augment wild populations: bustards

	• Release captive-bred individuals into the wild to restore or augment wild populations: gamebirds • Release captive-bred individuals into the wild to restore or augment wild populations: owls • Release captive-bred individuals into the wild restore or augment wild populations: parrots • Release captive-bred individuals into the wild to restore or augment wild populations: pigeons • Release captive-bred individuals into the wild to restore or augment wild populations: rails • Release captive-bred individuals into the wild to restore or augment wild populations: storks and ibises • Release captive-bred individuals into the wild to restore or augment wild populations: waders • Release captive-bred individuals into the wild to restore or augment wild populations: wildfowl • Release chicks and adults in 'coveys' • Use 'anti-predator training' to improve survival after release • Use appropriate populations to source released populations • Use 'flying training' before release • Use holding pens at release sites • Use microlites to help birds migrate

Likely to be beneficial

Provide supplementary food after release

All three studies captured found that released birds used supplementary food provided. One study from Australia found that malleefowl had higher survival when provided with food and a study from Peru found that supplementary food could be used to increase the foraging ranges of Andean condors after release. *Assessment: likely to be beneficial (effectiveness 45%; certainty 48%; harms 0%).*

http://www.conservationevidence.com/actions/639

Release captive-bred individuals into the wild to restore or augment wild populations (cranes)

Four studies of five release programmes from the USA and Russia found that released cranes had high survival or bred in the wild. Two studies from two release programmes in the USA found low survival of captive-bred eggs fostered to wild birds compared with wild eggs, or a failure to increase the wild flock size. A worldwide review found that releases of migratory species were more successful if birds were released into existing flocks, and for non-migratory populations. One study from the USA found that birds released as sub-adults had higher survival than birds cross-fostered to wild birds. *Assessment: likely to be beneficial (effectiveness 55%; certainty 50%; harms 5%).*

http://www.conservationevidence.com/actions/621

Release captive-bred individuals into the wild to restore or augment wild populations (raptors)

Five studies of three release programmes from across the world found the establishment or increase of wild populations of falcons. Five studies from the USA found high survival of released raptors although one study from Australia found that a wedge-tailed eagle had to be taken back into captivity after acting aggressively towards humans, and another Australian study found that only one of 15 brown goshawks released was recovered. *Assessment: likely to be beneficial (effectiveness 69%; certainty 56%; harms 10%).*

http://www.conservationevidence.com/actions/626

Release captive-bred individuals into the wild to restore or augment wild populations (songbirds)

A study in Mauritius describes the establishment of a population of Mauritius fody following the release of captive-bred individuals. Four studies of three release programmes on Hawaii found high survival of all three species released, with two thrush species successfully breeding. A replicated, controlled study from the USA found that shrike pairs with captive-bred females had lower reproductive success than pairs where both parents were wild-bred. *Assessment: likely to be beneficial (effectiveness 42%; certainty 40%; harms 5%).*

http://www.conservationevidence.com/actions/630

Release captive-bred individuals into the wild to restore or augment wild populations (vultures)

Four studies of two release programmes found that release programmes led to large population increases in Andean condors in Colombia and griffon vultures in France. A small study in Peru found high survival of released Andean condors over 18 months, with all fatalities occurring in the first six months after release. *Assessment: likely to be beneficial (effectiveness 73%; certainty 54%; harms 0%).*

http://www.conservationevidence.com/actions/625

Unknown effectiveness (limited evidence)

Clip birds' wings on release

Two of four studies found that bustards and geese had lower survival when released into holding pens with clipped wings compared to birds released without clipped wings. One study found no differences in survival for clipped or unclipped northern bald ibis. One study found that adult geese released with clipped wings survived better than geese released before they were able to fly. *Assessment: unknown effectiveness — limited evidence (effectiveness 10%; certainty 30%; harms 5%).*

http://www.conservationevidence.com/actions/633

Release birds as adults or sub-adults not juveniles

Three out of nine studies from across the world found that birds released as sub-adults had higher survival than those released as juveniles. Two studies found lower survival of wing-clipped sub-adult geese and bustards, compared with juveniles and one study found lower survival of all birds released as sub-adults, compared to those released as juveniles. Three studies found no differences in survival for birds released at different ages, although one found higher reproduction in birds released at greater ages. *Assessment: unknown effectiveness — limited evidence (effectiveness 35%; certainty 15%; harms 19%).*

http://www.conservationevidence.com/actions/636

Release birds in groups

A study from New Zealand found that released stilts were more likely to move long distances after release if they were released in larger groups. *Assessment: unknown effectiveness — limited evidence (effectiveness 32%; certainty 26%; harms 2%).*

http://www.conservationevidence.com/actions/634

Release captive-bred individuals into the wild to restore or augment wild populations (bustards)

Three reviews of a release programme for houbara bustard in Saudi Arabia found low initial survival of released birds, but the establishment of a breeding population and an overall success rate of 41%. The programme tested many different release techniques, the most successful of which was release of sub-adults, which were able to fly, into a large exclosure. *Assessment: unknown effectiveness — limited evidence (effectiveness 34%; certainty 26%; harms 5%).*

http://www.conservationevidence.com/actions/622

Release captive-bred individuals into the wild to restore or augment wild populations (gamebirds)

One of five studies from across the world found that releasing gamebirds established a population or bolstered an existing population. A review of a reintroduction programme in Pakistan found some breeding success in released cheer pheasants, but habitat change at the release site then excluded released birds. Three studies from Europe and the USA found that released birds had low survival, low reproductive success and no impact on the wild population. *Assessment: unknown effectiveness — limited evidence (effectiveness 5%; certainty 35%; harms 1%).*

http://www.conservationevidence.com/actions/619

Release captive-bred individuals into the wild to restore or augment wild populations (owls)

A study in the USA found that a barn owl population was established following the release of 157 birds in the area over three years. A replicated, controlled study in Canada found that released burrowing owls had similar reproductive output but higher mortality than wild birds. *Assessment: unknown effectiveness — limited evidence (effectiveness 24%; certainty 15%; harms 0%).*

http://www.conservationevidence.com/actions/627

Release captive-bred individuals into the wild to restore or augment wild populations (parrots)

A study from Venezuela found that the population of yellow-shouldered amazons increased significantly following the release of captive-bred birds along with other interventions. A study in Costa Rica and Peru found high survival and some breeding of scarlet macaw after release. Three replicated studies in the USA, Dominican Republic and Puerto Rico found low survival in released birds, although the Puerto Rican study also found that released birds bred successfully. *Assessment: unknown effectiveness — limited evidence (effectiveness 50%; certainty 30%; harms 3%).*

http://www.conservationevidence.com/actions/629

Release captive-bred individuals into the wild to restore or augment wild populations (pigeons)

A single review of a captive-release programme in Mauritius found that that released pink pigeons had a first year survival of 36%. *Assessment: unknown effectiveness — limited evidence (effectiveness 20%; certainty 5%; harms 1%).*

http://www.conservationevidence.com/actions/628

Release captive-bred individuals into the wild to restore or augment wild populations (rails)

One study from Australia found that released Lord Howe Island woodhens successfully bred in the wild, re-establishing a wild population and a study from the UK found high survival of released corncrake in the first summer after release. A replicated study in New Zealand found very low survival of North Island weka following release, mainly due to predation. *Assessment: unknown effectiveness — limited evidence (effectiveness 26%; certainty 16%; harms 0%).*

http://www.conservationevidence.com/actions/620

Release captive-bred individuals into the wild to restore or augment wild populations (storks and ibises)

A replicated study and a review of northern bald ibis release programmes in Europe and the Middle East found that only one of four resulted in a wild population being established or supported, with many birds dying or dispersing, rather than forming stable colonies. *Assessment: unknown effectiveness — limited evidence (effectiveness 20%; certainty 20%; harms 2%).*

http://www.conservationevidence.com/actions/624

Release captive-bred individuals into the wild to restore or augment wild populations (waders)

A review of black stilt releases in New Zealand found that birds had low survival (13–20%) and many moved away from their release sites. *Assessment: unknown effectiveness — limited evidence (effectiveness 10%; certainty 5%; harms 15%).*

http://www.conservationevidence.com/actions/623

Release captive-bred individuals into the wild to restore or augment wild populations (wildfowl)

Two studies of reintroduction programmes of ducks in New Zealand found high survival of released birds and population establishment. A study from Alaska found low survival of released cackling geese, but the population recovered from 1,000 to 6,000 birds after releases and the control of mammalian predators. A review of a reintroduction programme from Hawaii found that the release of Hawaiian geese (nene) did not result in the establishment of a self-sustaining population. Two studies from Canada found very low return rates for released ducks with one finding no evidence for survival of released birds over two years, although there was some evidence that breeding success was higher for released birds than wild ones. *Assessment: unknown effectiveness — limited evidence (effectiveness 30%; certainty 24%; harms 0%).*

http://www.conservationevidence.com/actions/618

Release chicks and adults in 'coveys'

Two out of three studies found that geese and partridges released in coveys had higher survival than young birds released on their own or adults released in pairs. A study from Saudi Arabia found that bustard chicks had low survival when released in coveys with flightless females. *Assessment: unknown effectiveness — limited evidence (effectiveness 40%; certainty 36%; harms 6%).*

http://www.conservationevidence.com/actions/635

Use 'anti-predator training' to improve survival after release

Both studies captured found higher survival for birds given predator training before release, compared with un-trained birds. One found that using a live fox, but not a model, for training increased survival in bustards, but that several birds were injured during training. *Assessment: unknown effectiveness — limited evidence (effectiveness 50%; certainty 20%; harms 9%).*

http://www.conservationevidence.com/actions/637

Use appropriate populations to source released populations

Two studies from Europe found that birds from populations near release sites adapted better and in one case had higher reproductive productivity than those from more distant populations. *Assessment: unknown effectiveness — limited evidence (effectiveness 53%; certainty 31%; harms 0%).*

http://www.conservationevidence.com/actions/631

Use 'flying training' before release

A study from the Dominican Republic found that parrots had higher first-year survival if they were given pre-release flying training. *Assessment: unknown effectiveness — limited evidence (effectiveness 30%; certainty 10%; harms 0%).*

http://www.conservationevidence.com/actions/638

Use holding pens at release sites

Three of four studies from North America and Saudi Arabia found that birds released into holding pens were more likely to form pairs or had higher survival than birds released into the open. One study found that parrots released into pens had lower survival than those released without preparation. A review of northern bald ibis releases found that holding pens could be used to prevent birds from migrating from the release site and so increase survival. *Assessment: unknown effectiveness — limited evidence (effectiveness 51%; certainty 36%; harms 2%).*

http://www.conservationevidence.com/actions/632

Use microlites to help birds migrate

A study from Europe found that northern bald ibises followed a microlite south in the winter but failed to make the return journey the next year. *Assessment: unknown effectiveness — limited evidence (effectiveness 3%; certainty 5%; harms 5%).*

http://www.conservationevidence.com/actions/640

4. FARMLAND CONSERVATION

Lynn V. Dicks, Joscelyne E. Ashpole, Juliana Dänhardt, Katy James, Annelie Jönsson, Nicola Randall, David A. Showler, Rebecca K. Smith, Susan Turpie, David R. Williams & William J. Sutherland

Expert assessors

Lynn V. Dicks, University of Cambridge, UK
Ian Hodge, University of Cambridge, UK
Clunie Keenleyside, Institute for European Environmental Policy, UK
Will Peach, Royal Society for the Protection of Birds, UK
Nicola Randall, Harper Adams University, UK
Jörn Scharlemann, United Nations Environment Programme — World Conservation Monitoring Centre, UK
Gavin Siriwardena, British Trust for Ornithology, UK
Henrik Smith, Lund University, Sweden
Rebecca K. Smith, University of Cambridge, UK
William J. Sutherland, University of Cambridge, UK

Scope of assessment: for native farmland wildlife in northern and western Europe (European countries west of Russia, but not south of France, Switzerland, Austria, Hungary and Romania).

Assessed: 2014.

Effectiveness measure is the % of experts that answered yes to the question: based on the evidence presented does this intervention benefit wildlife? (Yes, no or don't know).

Certainty measure is the median % score for the question: how much do we understand the extent to which this intervention benefits wildlife on farmland? (0 = no evidence, 100% = certainty).

Harm measure was not scored for this synopsis.

 http://dx.doi.org/10.11647/OBP.0109.04

This book is meant as a guide to the evidence available for different conservation interventions and as a starting point in assessing their effectiveness. The assessments are based on the available evidence for the target group of species for each intervention. The assessment may therefore refer to different species or habitat to the one(s) you are considering. Before making any decisions about implementing interventions it is vital that you read the more detailed accounts of the evidence in order to assess their relevance for your study species or system.

Full details of the evidence are available at
www.conservationevidence.com

There may also be significant negative side-effects on the target groups or other species or communities that have not been identified in this assessment.

A lack of evidence means that we have been unable to assess whether or not an intervention is effective or has any harmful impacts.

4.1 All farming systems

<table>
<tr><th colspan="2">Based on the collated evidence, what is the current assessment of the effectiveness of interventions for all farming systems?</th></tr>
<tr><td>Beneficial</td><td>• Create uncultivated margins around intensive arable or pasture fields
• Plant grass buffer strips/margins around arable or pasture fields
• Plant nectar flower mixture/wildflower strips
• Plant wild bird seed or cover mixture
• Provide or retain set-aside areas in farmland</td></tr>
<tr><td>Likely to be beneficial</td><td>• Manage ditches to benefit wildlife
• Manage hedgerows to benefit wildlife (includes no spray, gap-filling and laying)
• Pay farmers to cover the costs of conservation measures
• Provide supplementary food for birds or mammals</td></tr>
<tr><td>Unknown effectiveness (limited evidence)</td><td>• Connect areas of natural or semi-natural habitat
• Increase the proportion of natural or semi-natural habitat in the farmed landscape
• Make direct payments per clutch for farmland birds
• Manage the agricultural landscape to enhance floral resources
• Mark bird nests during harvest or mowing
• Plant new hedges
• Provide nest boxes for bees (solitary bees or bumblebees)
• Provide nest boxes for birds
• Provide other resources for birds (water, sand for bathing)
• Provide refuges during harvest or mowing</td></tr>
</table>

No evidence found (no assessment)	• Apply 'cross compliance' environmental standards linked to all subsidy payments • Implement food labelling schemes relating to biodiversity-friendly farming (organic, LEAF marque) • Introduce nest boxes stocked with solitary bees • Maintain in-field elements such as field islands and rockpiles • Manage stone-faced hedge banks to benefit wildlife • Manage woodland edges to benefit wildlife • Plant in-field trees (not farm woodland) • Protect in-field trees (includes management such as pollarding and surgery) • Provide badger gates • Provide foraging perches (e.g. for shrikes) • Provide otter holts • Provide red squirrel feeders • Reduce field size (or maintain small fields) • Restore or maintain dry stone walls • Support or maintain low-intensity agricultural systems

Beneficial

Create uncultivated margins around intensive arable or pasture fields

Twenty studies (including one randomized, replicated, controlled trial) from seven countries found uncultivated margins support more invertebrates, small mammal species or higher plant diversity than other habitats. Four studies (including two replicated studies from the UK) found positive associations between birds and uncultivated margins. Fifteen studies (including one randomized, replicated, controlled trial) from four countries found naturally regenerated margins had lower invertebrate or plant abundance or diversity than conventional fields or sown margins. Six studies (one randomized, replicated, controlled) from three countries found uncultivated margins did not have higher plant or invertebrate abundance or diversity than cropped or sown margins. *Assessment: beneficial (effectiveness 100%; certainty 63%).*

http://www.conservationevidence.com/actions/63

Plant grass buffer strips/margins around arable or pasture fields

Twenty studies (including two randomized, replicated, controlled studies) from four countries found grass margins benefited invertebrates, including increases in abundance or diversity. Nine studies (including two replicated, controlled trials) from the UK found grass buffer strips benefit birds, with increased numbers, diversity or use. Seven replicated studies (four controlled, two randomized) from two countries found grass buffer strips increased plant cover and species richness, a review found benefits to plants. Five studies (two replicated, controlled) from two countries found benefits to small mammals. Six (including three replicated, controlled trials) from two countries found no clear effect on invertebrate or bird numbers. *Assessment: beneficial (effectiveness 90%; certainty 65%).*

http://www.conservationevidence.com/actions/246

Plant nectar flower mixture/wildflower strips

Forty-one studies (including one randomized, replicated, controlled trial) from eight countries found flower strips increased invertebrate numbers or diversity. Ten studies (two replicated, controlled) found invertebrates visited flower strips. Fifteen studies (two randomized, replicated, controlled) found mixed or negative effects on invertebrates. Seventeen studies (one randomized, replicated, controlled) from seven countries found more plants or plant species on flower strips, four did not. Five studies (two randomized, replicated, controlled) from two countries found bird numbers, diversity or use increased in flower strips, two studies did not. Five studies (four replicated) found increases in small mammal abundance or diversity in flower strips. *Assessment: beneficial (effectiveness 100%; certainty 75%).*

http://www.conservationevidence.com/actions/442

Plant wild bird seed or cover mixture

Fifteen studies (including a systematic review) from the UK found fields sown with wild bird cover mix had more birds or bird species than other farmland habitats. Six studies (including two replicated trials) from the UK found birds used wild bird cover more than other habitats. Nine replicated studies from France and the UK found mixed or negative effects on birds. Eight studies (including two randomized, replicated, controlled studies) from the UK found wild bird cover had more invertebrates, four (including two replicated trials) found mixed or negative effects on invertebrate numbers. Six studies (including two replicated, controlled trials) from the UK found wild bird cover mix benefited plants, two replicated studies did not. *Assessment: beneficial (effectiveness 100%; certainty 65%).*

http://www.conservationevidence.com/actions/594

Provide or retain set-aside areas in farmland

Thirty-seven studies (one systematic review, no randomized, replicated, controlled trials) compared use of set-aside areas with control farmed fields. Twenty-one (including the systematic review) showed benefits to, or higher use by, all wildlife groups considered. Thirteen studies found some species or groups used set-aside more than crops; others did not. Two found higher Eurasian skylark reproductive success and one study found lower success on set-aside than control fields. Four studies found set-aside had no effect on wildlife, one found an adverse effect. Two studies found neither insects nor small mammals preferred set-aside. *Assessment: beneficial (effectiveness 90%; certainty 70%).*

http://www.conservationevidence.com/actions/156

Likely to be beneficial

Manage ditches to benefit wildlife

Five studies (including one replicated, controlled study) from the UK and the Netherlands found ditch management had positive effects on numbers, diversity or biomass of some or all invertebrates, amphibians, birds or plants studied. Three studies from the Netherlands and the UK (including two replicated site comparisons) found negative or no clear effects on plants or some birds. *Assessment: likely to be beneficial (effectiveness 40%; certainty 45%).*

http://www.conservationevidence.com/actions/135

Manage hedgerows to benefit wildlife (includes no spray, gap-filling and laying)

Ten studies from the UK and Switzerland (including one randomized, replicated, controlled trial) found managing hedges for wildlife increased berry yields, diversity or abundance of plants, invertebrates or birds. Five UK studies (including one randomized, replicated, controlled trial) found plants, bees and farmland birds were unaffected by hedge management. Two replicated studies found hedge management had mixed effects on invertebrates or reduced hawthorn berry yield. *Assessment: likely to be beneficial (effectiveness 70%; certainty 50%).*

http://www.conservationevidence.com/actions/116

Pay farmers to cover the cost of conservation measures (as in agri-environment schemes)

For birds, twenty-four studies (including one systematic review) found increases or more favourable trends in bird populations, while eleven studies (including one systematic review) found negative or no effects of agri-environment schemes. For plants, three studies found more plant species, two found fewer plant species and seven found little or no effect of agri-environment schemes. For invertebrates, five studies found increases in abundance or species richness, while six studies found little or no effect of agri-environment schemes. For mammals, one replicated study found positive effects of agri-environment schemes and three studies found mixed effects in different regions or for different species. *Assessment: likely to be beneficial (effectiveness 60%; certainty 50%).*

http://www.conservationevidence.com/actions/700

Provide supplementary food for birds or mammals

Nine studies (two randomized, replicated, controlled) from France, Sweden and the UK found providing supplementary food increased abundance, overwinter survival or productivity of some birds. Two of the studies did not separate the effects of several interventions. Four studies (one replicated, controlled and one randomized, replicated) from Finland and the UK found some birds or mammals used supplementary food. Six replicated studies (three controlled) from Sweden and the UK found no clear effect on some birds or plants. *Assessment: likely to be beneficial (effectiveness 90%; certainty 50%).*

http://www.conservationevidence.com/actions/648

Unknown effectiveness (limited evidence)

Connect areas of natural or semi-natural habitat

All four studies (including two replicated trials) from the Czech Republic, Germany and the Netherlands investigating the effects of linking patches of natural or semi-natural habitat found some colonization by invertebrates or mammals. Colonization by invertebrates was slow or its extent varied between taxa. *Assessment: unknown effectiveness — limited evidence (effectiveness 0%; certainty 15%).*

http://www.conservationevidence.com/actions/579

Increase the proportion of semi-natural habitat in the farmed landscape

Of five studies monitoring the effects of the Swiss Ecological Compensation Areas scheme at a landscape scale (including three replicated site comparisons), one found an increase in numbers of birds of some species, two found no effect on birds and three found some species or groups increasing and others decreasing. *Assessment: unknown effectiveness — limited evidence (effectiveness 20%; certainty 20%).*

http://www.conservationevidence.com/actions/145

Make direct payments per clutch for farmland birds

Two replicated, controlled studies from the Netherlands found per clutch payments did not increase overall bird numbers. A replicated site comparison from the Netherlands found more birds bred on 12.5 ha plots under management including per-clutch payments but there were no differences at the field-scale. *Assessment: unknown effectiveness — limited evidence (effectiveness 0%; certainty 20%).*

http://www.conservationevidence.com/actions/146

Manage the agricultural landscape to enhance floral resources

A large replicated, controlled study from the UK found the number of long-tongued bumblebees on field margins was positively correlated with the number of 'pollen and nectar' agri-environment agreements in a 10 km square. *Assessment: unknown effectiveness — limited evidence (effectiveness 40%; certainty 10%).*

http://www.conservationevidence.com/actions/362

Mark bird nests during harvest or mowing

A replicated study from the Netherlands found that marked northern lapwing nests were less likely to fail as a result of farming operations than unmarked nests. *Assessment: unknown effectiveness — limited evidence (effectiveness 20%; certainty 15%).*

http://www.conservationevidence.com/actions/148

Plant new hedges

Two studies (including one replicated trial) from France and the UK found new hedges had more invertebrates or plant species than fields or field margins. A review found new hedges had more ground beetles than older hedges. However, an unreplicated site comparison from Germany found only two out of 85 ground beetle species dispersed along new hedges. A review found lower pest outbreaks in areas with new hedges. *Assessment: unknown effectiveness — limited evidence (effectiveness 60%; certainty 25%).*

http://www.conservationevidence.com/actions/538

Provide nest boxes for bees (solitary bees or bumblebees)

Ten studies (nine replicated) from Germany, Poland and the UK found solitary bee nest boxes were used by bees. Two replicated trials from the UK found bumblebee nest boxes had very low uptake. Two replicated studies found the local population size or number of emerging red mason bees increased when nest boxes were provided. A replicated trial in Germany found the number of occupied solitary bee nests almost doubled over three years with repeated nest box provision. *Assessment: unknown effectiveness — limited evidence (effectiveness 90%; certainty 38%).*

http://www.conservationevidence.com/actions/80

Provide nest boxes for birds

Two studies (including one before-and-after trial) from the Netherlands and the UK found providing nest boxes increased the number of clutches or breeding adults of two bird species. A replicated study from Switzerland found nest boxes had mixed effects on the number of broods produced by two species. Eight studies (six replicated) from five countries found nest boxes were used by birds. A controlled study from the UK found one species did not use artificial nest sites. Three replicated studies (one paired) from the UK and Sweden found box location influenced use or nesting success. *Assessment: unknown effectiveness — limited evidence (effectiveness 30%; certainty 23%).*

http://www.conservationevidence.com/actions/155

Provide other resources for birds (water, sand for bathing)

A small study in France found grey partridge density was higher in areas where water, shelter, sand and food were provided. *Assessment: unknown effectiveness — limited evidence (effectiveness 0%; certainty 1%).*

http://www.conservationevidence.com/actions/117

Provide refuges during harvest or mowing

A replicated study from France found mowing refuges reduced contact between mowing machinery and unfledged quails and corncrakes. A replicated controlled study and a review from the UK found Eurasian skylark did not use nesting refuges more than other areas. *Assessment: unknown effectiveness — limited evidence (effectiveness 20%; certainty 11%).*

http://www.conservationevidence.com/actions/147

No evidence found (no assessment)

We have captured no evidence for the following interventions:

- Apply 'cross compliance' environmental standards linked to all subsidy payments
- Implement food labelling schemes relating to biodiversity-friendly farming (organic, LEAF marque)
- Introduce nest boxes stocked with solitary bees
- Maintain in-field elements such as field islands and rockpiles
- Manage stone-faced hedge banks to benefit wildlife
- Manage woodland edges to benefit wildlife
- Plant in-field trees (not farm woodland)

- Protect in-field trees (includes management such as pollarding and surgery)
- Provide badger gates
- Provide foraging perches (e.g. for shrikes)
- Provide otter holts
- Provide red squirrel feeders
- Reduce field size (or maintain small fields)
- Restore or maintain dry stone walls
- Support or maintain low intensity agricultural systems

4.2 Arable farming

Based on the collated evidence, what is the current assessment of the effectiveness of interventions for arable farming systems?	
Beneficial	• Create skylark plots • Leave cultivated, uncropped margins or plots (includes 'lapwing plots')
Likely to be beneficial	• Create beetle banks • Leave overwinter stubbles • Reduce tillage • Undersow spring cereals, with clover for example
Unknown effectiveness (limited evidence)	• Convert or revert arable land to permanent grassland • Create rotational grass or clover leys • Increase crop diversity • Plant cereals in wide-spaced rows • Plant crops in spring rather than autumn • Plant nettle strips • Sow rare or declining arable weeds
No evidence found (no assessment)	• Add 1% barley into wheat crop for corn buntings • Create corn bunting plots • Leave unharvested cereal headlands within arable fields • Use new crop types to benefit wildlife (such as perennial cereal crops)
Evidence not assessed	• Implement 'mosaic management', a Dutch agri-environment option • Plant more than one crop per field (intercropping) • Take field corners out of management

Beneficial

Create skylark plots

All four studies (two replicated, controlled trials) from Switzerland and the UK investigating the effect of skylark plots on Eurasian skylarks found positive effects, including increases in population size. A replicated study from Denmark found skylarks used undrilled patches in cereal fields. Three studies (one replicated, controlled) from the UK found benefits to plants and invertebrates. Two replicated studies (one controlled) from the UK found no significant differences in numbers of invertebrates or seed-eating songbirds. *Assessment: beneficial (effectiveness 100%; certainty 80%).*

http://www.conservationevidence.com/actions/540

Leave cultivated, uncropped margins or plots (includes 'lapwing plots')

Seventeen of nineteen individual studies looking at uncropped, cultivated margins or plots (including one replicated, randomized, controlled trial) primarily from the UK found benefits to some or all target farmland bird species, plants, invertebrates or mammals. Two studies (one replicated) from the UK found no effect on ground beetles or most farmland birds. Two replicated site comparisons from the UK found cultivated, uncropped margins were associated with lower numbers of some bird species or age groups in some areas. *Assessment: beneficial (effectiveness 100%; certainty 65%).*

http://www.conservationevidence.com/actions/562

Likely to be beneficial

Create beetle banks

Five reports from two replicated studies (one controlled) and a review from Denmark and the UK found beetle banks had positive effects on invertebrate numbers, diversity or distributions. Five replicated studies (two controlled) found lower or no difference in invertebrate numbers. Three studies (including a replicated, controlled trial) from the UK found beetle banks, alongside other management, had positive effects on bird numbers or usage. Three studies (one replicated site comparison) from the

UK found mixed or no effects on birds, two found negative on no clear effects on plants. Two studies (one controlled) from the UK found harvest mice nested on beetle banks. *Assessment: likely to be beneficial (effectiveness 80%; certainty 60%).*

http://www.conservationevidence.com/actions/651

Leave overwinter stubbles

Eighteen studies investigated the effects of overwinter stubbles. Thirteen studies (including two replicated site comparisons and a systematic review) from Finland, Switzerland and the UK found leaving overwinter stubbles benefits some plants, invertebrates, mammals or birds. Three UK studies (one randomized, replicated, controlled) found only certain birds were positively associated with overwinter stubbles. *Assessment: likely to be beneficial (effectiveness 90%; certainty 50%).*

http://www.conservationevidence.com/actions/695

Reduce tillage

Thirty-four studies (including seven randomized, replicated, controlled trials) from nine countries found reducing tillage had some positive effects on invertebrates, weeds or birds. Twenty-seven studies (including three randomized, replicated, controlled trials) from nine countries found reducing tillage had negative or no clear effects on some invertebrates, plants, mammals or birds. Three of the studies did not distinguish between the effects of reducing tillage and reducing chemical inputs. *Assessment: likely to be beneficial (effectiveness 60%; certainty 60%).*

http://www.conservationevidence.com/actions/126

Undersow spring cereals, with clover for example

Eleven studies (including three randomized, replicated, controlled trials) from Denmark, Finland, Switzerland and the UK found undersowing spring cereals benefited some birds, plants or invertebrates, including increases in numbers or species richness. Five studies (including one replicated, randomized, controlled trial) from Austria, Finland and the UK found no benefits to invertebrates, plants or some birds. *Assessment: likely to be beneficial (effectiveness 60%; certainty 43%).*

http://www.conservationevidence.com/actions/136

Unknown effectiveness (limited evidence)

Convert or revert arable land to permanent grassland

All seven individual studies (including two replicated, controlled trials) from the Czech Republic, Denmark and the UK looking at the effects of reverting arable land to grassland found no clear benefits to birds, mammals or plants. *Assessment: unknown effectiveness — limited evidence (effectiveness 0%; certainty 20%).*

http://www.conservationevidence.com/actions/561

Create rotational grass or clover leys

A controlled study from Finland found more spiders and fewer pest insects in clover leys than the crop. A replicated study from the UK found grass leys had fewer plant species than other conservation habitats. A UK study found newer leys had lower earthworm abundance and species richness than older leys. *Assessment: unknown effectiveness — limited evidence (effectiveness 0%; certainty 10%).*

http://www.conservationevidence.com/actions/643

Increase crop diversity

Four studies (including one replicated, controlled trial) from Belgium, Germany and Hungary found more ground beetle or plant species or individuals in fields with crop rotations or on farms with more crops in rotation than monoculture fields. *Assessment: unknown effectiveness — limited evidence (effectiveness 0%; certainty 9%).*

http://www.conservationevidence.com/actions/560

Plant cereals in wide-spaced rows

Two studies (one randomized, replicated, controlled) from the UK found planting cereals in wide-spaced rows had inconsistent, negative or no effects on plant and invertebrate abundance or species richness. *Assessment: unknown effectiveness — limited evidence (effectiveness 0%; certainty 18%).*

http://www.conservationevidence.com/actions/564

Plant crops in spring rather than autumn

Seven studies (including two replicated, controlled trials) from Denmark, Sweden and the UK found sowing crops in spring had positive effects on farmland bird numbers or nesting rates, invertebrate numbers or weed diversity or density. Three of the studies found the effects were seasonal. A review of European studies found fewer invertebrates in spring wheat than winter wheat. *Assessment: unknown effectiveness — limited evidence (effectiveness 40%; certainty 35%).*

http://www.conservationevidence.com/actions/137

Plant nettle strips

A small study from Belgium found nettle strips in field margins had more predatory invertebrate species than the crop, but fewer individuals than the crop or natural nettle stands. *Assessment: unknown effectiveness — limited evidence (effectiveness 50%; certainty 10%).*

http://www.conservationevidence.com/actions/118

Sow rare or declining arable weeds

Two randomized, replicated, controlled studies from the UK identified factors important in establishing rare or declining arable weeds, including type of cover crop, cultivation and herbicide treatment. *Assessment: unknown effectiveness — limited evidence (effectiveness 40%; certainty 15%).*

http://www.conservationevidence.com/actions/642

No evidence found (no assessment)

We have captured no evidence for the following interventions:

- Add 1% barley into wheat crop for corn buntings
- Create corn bunting plots
- Leave unharvested cereal headlands in arable fields
- Use new crop types to benefit wildlife (such as perennial cereal crops)

Evidence not assessed

Implement 'mosaic management', a Dutch agri-environment option

A replicated, controlled before-and-after study from the Netherlands found mosaic management had mixed effects on population trends of wading bird species. A replicated, paired sites study from the Netherlands found one bird species had higher productivity under mosaic management. *Assessment: this intervention has not been assessed.*

http://www.conservationevidence.com/actions/130

Plant more than one crop per field (intercropping)

All five studies (including three randomized, replicated, controlled trials) from the Netherlands, Poland, Switzerland and the UK looking at the effects of planting more than one crop per field found increases in the number of earthworms or ground beetles. *Assessment: this intervention has not been assessed.*

http://www.conservationevidence.com/actions/124

Take field corners out of management

A replicated site comparison from the UK found a positive correlation between grey partridge overwinter survival and taking field corners out of management. Brood size, ratio of young to old birds and density changes were unaffected. *Assessment: this intervention has not been assessed.*

http://www.conservationevidence.com/actions/128

4.3 Perennial (non-timber) crops

Based on the collated evidence, what is the current assessment of the effectiveness of interventions for perennial (non-timber) crops?	
Unknown effectiveness (limited evidence)	• Maintain traditional orchards
No evidence found (no assessment)	• Manage short-rotation coppice to benefit wildlife (includes 8m rides) • Restore or create traditional orchards

Unknown effectiveness (limited evidence)

Maintain traditional orchards

A replicated, controlled site comparison from Germany found more plant species in mown orchards than grazed or abandoned ones, but found no effects on wasps or bees. Two replicated site comparisons from Germany and Switzerland found traditional orchards managed under agri-environment schemes either did not have more plant species than controls or offered no clear benefits to birds. *Assessment: unknown effectiveness — limited evidence (effectiveness 10%; certainty 15%).*

http://www.conservationevidence.com/actions/703

No evidence found (no assessment)

We have captured no evidence for the following interventions:

- Manage short-rotation coppice to benefit wildlife (includes 8 m rides)
- Restore or create traditional orchards

4.4 Livestock farming

Based on the collated evidence, what is the current assessment of the effectiveness of interventions for livestock farming?	
Beneficial	• Restore or create species-rich semi-natural grassland • Use mowing techniques to reduce mortality
Likely to be beneficial	• Delay mowing or first grazing date on grasslands • Leave uncut strips of rye grass on silage fields • Maintain species-rich, semi-natural grassland • Maintain traditional water meadows (includes management for breeding and/or wintering waders/waterfowl) • Maintain upland heath/moorland • Reduce management intensity on permanent grasslands (several interventions at once) • Restore or create traditional water meadows
Unknown effectiveness (limited evidence)	• Add yellow rattle seed *Rhinanthus minor* to hay meadows • Employ areas of semi-natural habitat for rough grazing (includes salt marsh, lowland heath, bog, fen) • Exclude livestock from semi-natural habitat (including woodland) • Maintain wood pasture and parkland • Plant cereals for whole crop silage • Raise mowing height on grasslands • Restore or create upland heath/moorland • Restore or create wood pasture • Use traditional breeds of livestock

Likely to be ineffective or harmful	• Reduce grazing intensity on grassland (including seasonal removal of livestock)
No evidence found (no assessment)	• Maintain rush pastures • Mark fencing to avoid bird mortality • Plant Brassica fodder crops (grazed *in situ*)
Evidence not assessed	• Create open patches or strips in permanent grassland • Provide short grass for birds • Use mixed stocking

Beneficial

● Restore or create species-rich, semi-natural grassland

Twenty studies (including three randomized, replicated, controlled trials) from six countries found restored species-rich, semi-natural grasslands had similar invertebrate, plant or bird diversity or abundance to other grasslands. Seven studies (two randomized, replicated, controlled trials) from five countries found no clear effect on plant or invertebrate numbers, three replicated studies (of which two site comparisons) from two countries found negative effects. Forty studies (including six randomized, replicated, controlled trials) from nine countries identified effective techniques for restoring species-rich grassland. *Assessment: beneficial (effectiveness 100%; certainty 73%).*

http://www.conservationevidence.com/actions/133

● Use mowing techniques to reduce mortality

Seven studies (including two replicated trials, one controlled and one randomized) from Germany, Ireland, Switzerland and the UK found mowing techniques that reduced mortality or injury in amphibians, birds, invertebrates or mammals. A review found the UK corncrake population increased around the same time that Corncrake Friendly Mowing was introduced and a replicated trial found mowing from the field centre outwards reduced corncrake chick mortality. *Assessment: beneficial (effectiveness 100%; certainty 78%).*

http://www.conservationevidence.com/actions/698

Likely to be beneficial

Delay mowing or first grazing date on grasslands

Eight studies (including a European systematic review) from the Netherlands, Sweden and the UK found delaying mowing or grazing benefited some or all plants, invertebrates or birds, including increases in numbers or productivity. Three reviews found the UK corncrake population increased following management that included delayed mowing. Six studies (including a European systematic review) from five countries found no clear effect on some plants, invertebrates or birds. *Assessment: likely to be beneficial (effectiveness 60%; certainty 45%).*

http://www.conservationevidence.com/actions/131

Leave uncut strips of rye grass on silage fields

Four studies (including two replicated, controlled trials) from the UK found uncut strips of rye grass benefited some birds, with increased numbers. A randomized, replicated, controlled study from the UK found higher ground beetle diversity on uncut silage plots, but only in the third study year. *Assessment: likely to be beneficial (effectiveness 80%; certainty 49%).*

http://www.conservationevidence.com/actions/132

Maintain species-rich, semi-natural grassland

Nine studies (including two randomized, replicated before-and-after trials) from Switzerland and the UK looked at the effectiveness of agri-environment schemes in maintaining species-rich grassland and all except one found mixed results. All twelve studies (including a systematic review) from six countries looking at grassland management options found techniques that improved or maintained vegetation quality. A site comparison from Finland and Russia found butterfly communities were more affected by grassland age and origin than present management. *Assessment: likely to be beneficial (effectiveness 80%; certainty 60%).*

http://www.conservationevidence.com/actions/702

Maintain traditional water meadows (includes management for breeding and/or wintering waders/waterfowl)

Four studies (including a replicated site comparison) from Belgium, Germany, the Netherlands and the UK found maintaining traditional water meadows increased numbers of some birds or plant diversity. One bird species declined. Two studies (including a replicated site comparison from the Netherlands) found mixed or inconclusive effects on birds, plants or wildlife generally. A replicated study from the UK found productivity of one wading bird was too low to sustain populations in some areas of wet grassland managed for wildlife. *Assessment: likely to be beneficial (effectiveness 56%; certainty 50%).*

http://www.conservationevidence.com/actions/696

Maintain upland heath/moorland

Eight studies (including one randomized, replicated, controlled trial) from the UK found management, including reducing grazing, can help to maintain the conservation value of upland heath or moorland. Benefits included increased numbers of plants or invertebrates. Three studies (including a before-and-after trial) from the UK found management to maintain upland heath or moorland had mixed effects on some wildlife groups. Four studies (including a controlled site comparison) from the UK found reducing grazing had negative impacts on soil organisms, but a randomized, replicated before-and-after study found heather cover declined where grazing intensity had increased. *Assessment: likely to be beneficial (effectiveness 90%; certainty 50%).*

http://www.conservationevidence.com/actions/647

Reduce management intensity on permanent grasslands (several interventions at once)

Eleven studies (including four replicated site comparisons) from three countries found reducing management intensity benefited plants. Sixteen studies (including four paired site comparisons) from four countries found benefits to some or all invertebrates. Five studies (including one paired, replicated site comparison) from four countries found positive effects on some or all birds. Twenty-one studies (including two randomized, replicated, controlled trials) from six countries found no clear effects of reducing management intensity on some or all plants, invertebrates or birds.

Five studies (including two paired site comparisons) from four countries found negative effects on plants, invertebrates or birds. *Assessment: likely to be beneficial (effectiveness 100%; certainty 60%).*

http://www.conservationevidence.com/actions/69

Restore or create traditional water meadows

Three studies (two before-and-after trials) from Sweden and the UK looked at bird numbers following water meadow restoration, one found increases, one found increases and decreases, one found no increases. Seventeen studies (two randomized, replicated, controlled) from six countries found successful techniques for restoring wet meadow plant communities. Three studies (one replicated, controlled) from four countries found restoration of wet meadow plant communities had reduced or limited success. *Assessment: likely to be beneficial (effectiveness 100%; certainty 50%).*

http://www.conservationevidence.com/actions/119

Unknown effectiveness (limited evidence)

Add yellow rattle seed *Rhinanthus minor* to hay meadows

A review from the UK reported that hay meadows had more plant species when yellow rattle was present. A randomized, replicated controlled trial in the UK found yellow rattle could be established by 'slot seeding'. *Assessment: unknown effectiveness — limited evidence (effectiveness 70%; certainty 20%).*

http://www.conservationevidence.com/actions/129

Employ areas of semi-natural habitat for rough grazing (includes salt marsh, lowland heath, bog, fen)

Three studies (two replicated) from the UK and unspecified European countries found grazing had positive effects on birds, butterflies or biodiversity generally. A series of site comparisons from the UK found one bird species used heathland managed for grazing as feeding but not nesting sites. Two studies (one replicated site comparison) from the UK found grazing had negative effects on two bird species. *Assessment: unknown effectiveness — limited evidence (effectiveness 20%; certainty 10%).*

http://www.conservationevidence.com/actions/697

Exclude livestock from semi-natural habitat (including woodland)

Three studies (including one randomized, replicated, controlled trial) from Ireland and the UK found excluding livestock from semi-natural habitats benefited plants and invertebrates. Three studies (one replicated, controlled and one replicated paired sites comparison) from Ireland and the UK did not find benefits to plants or birds. Two studies (one replicated, controlled and a review) from Poland and the UK found limited or mixed effects. *Assessment: unknown effectiveness — limited evidence (effectiveness 20%; certainty 15%).*

http://www.conservationevidence.com/actions/150

Maintain wood pasture and parkland

A randomized, replicated, controlled trial in Sweden found annual mowing on wood pasture maintained the highest number of plant species. *Assessment: unknown effectiveness — limited evidence (effectiveness 40%; certainty 10%).*

http://www.conservationevidence.com/actions/649

Plant cereals for whole crop silage

A replicated study from the UK found cereal-based whole crop silage had higher numbers of some birds than other crops. A review from the UK reported that seed-eating birds avoided cereal-based whole crop silage in winter, but used it as much as spring barley in summer. *Assessment: unknown effectiveness — limited evidence (effectiveness 80%; certainty 28%).*

http://www.conservationevidence.com/actions/149

Raise mowing height on grasslands

Three studies (including one replicated, controlled trial) from the UK or unspecified European countries found raised mowing heights caused less damage to amphibians and invertebrates or increased Eurasian skylark productivity. Two studies (one randomized, replicated, controlled) from the UK found no effect on bird or invertebrate numbers and a replicated study from the UK found young birds had greater foraging success in shorter grass. *Assessment: unknown effectiveness — limited evidence (effectiveness 0%; certainty 35%).*

http://www.conservationevidence.com/actions/138

Restore or create upland heath/moorland

A small trial in northern England found moorland restoration increased the number of breeding northern lapwing. A UK review concluded that vegetation changes were slow during the restoration of heather moorland from upland grassland. *Assessment: unknown effectiveness — limited evidence (effectiveness 78%; certainty 20%).*

http://www.conservationevidence.com/actions/650

Restore or create wood pasture

A replicated, controlled trial in Belgium found survival and growth of tree seedlings planted in pasture was enhanced when they were protected from grazing. A replicated study in Switzerland found cattle browsing had negative effects on tree saplings. *Assessment: unknown effectiveness — limited evidence (effectiveness 40%; certainty 5%).*

http://www.conservationevidence.com/actions/644

Use traditional breeds of livestock

Three studies (one replicated) from the UK found the breed of livestock affected vegetation structure, invertebrate communities and the amount of plants grazed. A replicated trial from France, Germany and the UK found no difference in the number of plant species or the abundance of birds, invertebrates or mammals between areas grazed by traditional or commercial livestock. *Assessment: unknown effectiveness — limited evidence (effectiveness 0%; certainty 20%).*

http://www.conservationevidence.com/actions/539

Likely to be ineffective or harmful

Reduce grazing intensity on grassland (including seasonal removal of livestock)

Fifteen studies (including three randomized, replicated, controlled trials) from four countries found reducing grazing intensity benefited birds, invertebrates or plants. Three studies (including one randomized, replicated, controlled trial) from the Netherlands and the UK found no benefit to plants or invertebrates. Nine studies (including a systematic review) from France, Germany and the UK found mixed effects for some or all wildlife groups. The systematic review concluded that intermediate grazing levels are usually optimal but different wildlife groups are likely to have different grazing requirements. *Assessment: likely to be ineffective (effectiveness 30%; certainty 70%).*

http://www.conservationevidence.com/actions/704

No evidence found (no assessment)

We have captured no evidence for the following interventions:

- Maintain rush pastures
- Mark fencing to avoid bird mortality
- Plant brassica fodder crops (grazed *in situ*)

Evidence not assessed

Create open patches or strips in permanent grassland

A randomized, replicated, controlled study from the UK found more Eurasian skylarks used fields containing open strips, but numbers varied. A randomized, replicated, controlled study from the UK found insect numbers on grassy headlands initially dropped when strips were cleared. *Assessment:this intervention has not been assessed.*

http://www.conservationevidence.com/actions/563

Provide short grass for birds

A replicated UK study found two bird species spent more time foraging on short grass than longer grass. *Assessment: this intervention has not been assessed.*

http://www.conservationevidence.com/actions/115

Use mixed stocking

A replicated, controlled study in the UK found more spiders, harvestmen and pseudoscorpions in grassland grazed by sheep-only than grassland grazed by sheep and cattle. Differences were only found when suction sampling not pitfall-trapping. *Assessment: this intervention has not been assessed.*

http://www.conservationevidence.com/actions/93

4.5 Threat: Residential and commercial development

Based on the collated evidence, what is the current assessment of the effectiveness of interventions for residential and commercial development?	
Unknown effectiveness (limited evidence)	• Provide owl nest boxes (tawny owl, barn owl)
No evidence found (no assessment)	• Maintain traditional farm buildings • Provide bat boxes, bat grilles, improvements to roosts

Unknown effectiveness (limited evidence)

Provide owl nest boxes (tawny owl, barn owl)

Two studies (one before-and-after study) from the Netherlands and the UK found providing nest boxes increased barn owl populations. A replicated study from the UK found a decrease in the proportion of breeding barn owls was not associated with the number of nest boxes. *Assessment: unknown effectiveness — limited evidence (effectiveness 100%; certainty 33%).*

http://www.conservationevidence.com/actions/154

No evidence found (no assessment)

We have captured no evidence for the following interventions:

- Maintain traditional farm buildings
- Provide bat boxes, bat grilles, improvements to roosts

4.6 Threat: Agri-chemicals

Based on the collated evidence, what is the current assessment of the effectiveness of interventions for agri-chemicals?	
Beneficial	• Leave headlands in fields unsprayed (conservation headlands) • Reduce fertilizer, pesticide or herbicide use generally • Use organic rather than mineral fertilisers
Likely to be beneficial	• Reduce chemical inputs in grassland management
Unknown effectiveness (limited evidence)	• Provide buffer strips alongside water courses (rivers and streams) • Restrict certain pesticides
No evidence found (no assessment)	• Buffer in-field ponds
Evidence not assessed	• Make selective use of spring herbicides

Beneficial

Leave headlands in fields unsprayed (conservation headlands)

Twenty-two studies from 14 experiments (including two randomized, replicated, controlled) from five countries found conservation headlands had higher invertebrate or plant diversity than other habitats, twelve studies from ten experiments (three randomized, replicated, controlled) did not. Twenty-seven studies from 15 experiments (of which 13 replicated, controlled) from five countries found positive effects on abundance or behaviour of some wildlife groups. Nineteen studies from 13 experiments

(12 replicated, controlled) from four countries found similar, or lower, numbers of birds, invertebrates or plants on conservation headlands than other habitats. *Assessment: beneficial (effectiveness 90%; certainty 75%).*

http://www.conservationevidence.com/actions/652

Reduce fertilizer, pesticide or herbicide use generally

Thirty-four studies (including a systematic review) from 10 countries found reducing fertilizer, pesticide or herbicide inputs benefited some invertebrates, plants or birds. Twenty-five studies (including seven randomized, replicated, controlled trials) from eight countries found negative or no clear effects on some invertebrates, plants or birds. *Assessment: beneficial (effectiveness 100%; certainty 70%).*

http://www.conservationevidence.com/actions/139

Use organic rather than mineral fertilizers

Fourteen studies (including four randomized, replicated, controlled trials) from six countries found areas treated with organic rather than mineral fertilizers had more plants or invertebrates or higher diversity. A randomized, replicated, controlled trial from the UK found no effect on weed numbers. Two studies (including a small trial from Belgium) found organic fertilizers benefited invertebrates, a UK review found that in large quantities they did not. *Assessment: beneficial (effectiveness 100%; certainty 70%).*

http://www.conservationevidence.com/actions/134

Likely to be beneficial

Reduce chemical inputs in grassland management

Six studies (including a randomized, replicated, controlled before-and-after trial) from three countries found stopping fertilizer inputs on grassland improved plant or invertebrate species richness or abundance. Two reviews from the Netherlands and the UK found no or low fertilizer input grasslands favour some birds and invertebrates. Five studies (two replicated trials of which one randomized and one replicated) from three countries found no clear effects on invertebrates or plants. *Assessment: likely to be beneficial (effectiveness 90%; certainty 60%).*

http://www.conservationevidence.com/actions/694

Unknown effectiveness (limited evidence)

Provide buffer strips alongside water courses (rivers and streams)

Three studies (including one replicated site comparison) from the Netherlands and the UK found riparian buffer strips increased diversity or abundance of plants, invertebrates or birds and supported vegetation associated with water vole habitats. Two replicated site comparisons from France and Ireland found farms with buffer strips did not have more plant species than farms without strips. *Assessment: unknown effectiveness — limited evidence (effectiveness 10%; certainty 15%).*

http://www.conservationevidence.com/actions/120

Restrict certain pesticides

A small UK study found two fungicides that reduced insect abundance less than an alternative. A replicated, controlled trial in Switzerland found applying slug pellets in a band at the field edge was as effective as spreading the pellets across the field. *Assessment: unknown effectiveness — limited evidence (effectiveness 50%; certainty 5%).*

http://www.conservationevidence.com/actions/565

No evidence found (no assessment)

We have captured no evidence for the following intervention:

- Buffer in-field ponds

Evidence not assessed

Make selective use of spring herbicides

A randomized, replicated, controlled study from the UK found spring herbicides had some benefits for beneficial weeds and arthropods. *Assessment: this intervention has not been assessed.*

http://www.conservationevidence.com/actions/98

4.7 Threat: Transport and service corridors

Based on the collated evidence, what is the current assessment of the effectiveness of interventions for transport and service corridors?	
No evidence found (no assessment)	• Manage land under power lines to benefit wildlife

No evidence found (no assessment)

We have captured no evidence for the following intervention:

- Manage land under power lines to benefit wildlife

4.8 Threat: Hunting and trapping (for pest control, food or sport)

Based on the collated evidence, what is the current assessment of the effectiveness of interventions for hunting and trapping (for pest control, food or sport)?	
Unknown effectiveness (limited evidence)	• Enforce legislation to protect birds against persecution • Provide 'sacrificial' grasslands to reduce the impact of wild geese on crops
No evidence found (no assessment)	• Avoid use of lead shot • Use alerts to reduce grey partridge by-catch during shoots
Evidence not assessed	• Use scaring devices (e.g. gas guns) and other deterrents to reduce persecution of native species

Unknown effectiveness (limited evidence)

Enforce legislation to protect birds against persecution

Two before-and-after studies from Denmark and the UK found increased numbers or survival of raptors under legislative protection. *Assessment: unknown effectiveness — limited evidence (effectiveness 90%; certainty 18%).*

http://www.conservationevidence.com/actions/101

Provide 'sacrificial' grasslands to reduce the impact of wild geese on crops

All six studies from the UK (including four replicated, controlled trials) found that managing grasslands for geese increased the number of geese using these areas. Four of these studies found geese were moving within the study sites. *Assessment: unknown effectiveness — limited evidence (effectiveness 20%; certainty 5%).*

http://www.conservationevidence.com/actions/641

No evidence found (no assessment)

We have captured no evidence for the following interventions:

- Avoid use of lead shot
- Use alerts to reduce grey partridge by-catch during shoots

Evidence not assessed

Use scaring devices (e.g. gas guns) and other deterrents to reduce persecution of native species

A replicated, controlled trial in Germany found phosphorescent tape was more effective than normal yellow tape at deterring one of three mammal species. *Assessment: this intervention has not been assessed.*

http://www.conservationevidence.com/actions/645

4.9 Threat: Natural system modification

Based on the collated evidence, what is the current assessment of the effectiveness of interventions for natural system modification?	
Likely to be beneficial	• Raise water levels in ditches or grassland
Unknown effectiveness (limited evidence)	• Create scrapes and pools • Manage heather by swiping to simulate burning • Mange heather, gorse or grass by burning • Remove flood defence banks to allow inundation
No evidence found (no assessment)	• Re-wet moorland

Likely to be beneficial

Raise water levels in ditches or grassland

Eight studies (including two replicated, controlled trials) from Denmark, the Netherlands and the UK found raising water levels increased numbers of birds, invertebrates or plants or allowed wet grassland plant species to establish more rapidly. Three studies (two replicated) from the Netherlands and the UK found raising water levels had negative, limited or no effects on plants or birds. A replicated study from the UK found unflooded pastures had a greater weight of soil invertebrates than flooded pastures. *Assessment: likely to be beneficial (effectiveness 100%; certainty 55%).*

http://www.conservationevidence.com/actions/121

Unknown effectiveness (limited evidence)

Create scrapes and pools

Five studies (including a replicated, controlled, paired trial) from Sweden and the UK found creating scrapes and pools provided habitat for birds, invertebrates or plants or increased invertebrate diversity. Two replicated studies (one controlled, paired) from Ireland and the UK found mixed or no differences in invertebrate numbers between created ponds and controls or natural ponds. A study in Sweden found fewer fish species in constructed than natural wetlands. *Assessment: unknown effectiveness — limited evidence (effectiveness 100%; certainty 28%).*

http://www.conservationevidence.com/actions/153

Manage heather by swiping to simulate burning

A replicated, controlled trial from the UK found heather moorland subject to flailing had fewer plant species than burned plots but more species than unflailed plots. *Assessment: unknown effectiveness — limited evidence (effectiveness 40%; certainty 9%).*

http://www.conservationevidence.com/actions/151

Manage heather, gorse or grass by burning

A long-term replicated, controlled trial in Switzerland found burning of chalk grassland did not increase the number of plant species. A replicated, controlled trial in the UK found more plant species on burned than unburned heather moorland. *Assessment: unknown effectiveness — limited evidence (effectiveness 10%; certainty 5%).*

http://www.conservationevidence.com/actions/152

Remove flood defence banks to allow inundation

A controlled before-and-after study from the UK found a stretch of river that was allowed to flood had more bird species and territories than a channelized section. A study from Belgium found flooding and mowing increased plant species richness in meadow plots. *Assessment: unknown effectiveness — limited evidence (effectiveness 80%; certainty 10%).*

http://www.conservationevidence.com/actions/122

No evidence found (no assessment)

We have captured no evidence for the following intervention:

- Re-wet moorland

4.10 Threat: Invasive and other problematic species

Based on the collated evidence, what is the current assessment of the effectiveness of interventions for invasive and other problematic species?	
Likely to be beneficial	• Control predatory mammals and birds (foxes, crows, stoats and weasels)
Unknown effectiveness (limited evidence)	• Control scrub • Control weeds without damaging other plants in conservation areas • Protect individual nests of ground-nesting birds
No evidence found (no assessment)	• Control grey squirrels • Erect predator-proof fencing around important breeding sites for waders • Manage wild deer numbers • Remove coarse fish
Evidence not assessed	• Control bracken • Control invasive non-native plants on farmland (such as Himalayan balsam, Japanese knotweed) • Control mink • Provide medicated grit for grouse

Likely to be beneficial

Control predatory mammals and birds (foxes, crows, stoats and weasels)

Eight studies (including a systematic review) from France and the UK found predator control (sometimes alongside other interventions) increased the abundance, population size or productivity of some birds. A randomized, replicated, controlled study from the UK did not. *Assessment: likely to be beneficial (effectiveness 90%; certainty 60%).*

http://www.conservationevidence.com/actions/699

Unknown effectiveness (limited evidence)

Control scrub

A replicated site comparison from the UK found the number of young grey partridge per adult was negatively associated with management that included scrub control. *Assessment: unknown effectiveness — limited evidence (effectiveness 0%; certainty 2%).*

http://www.conservationevidence.com/actions/127

Control weeds without damaging other plants in conservation areas

Two studies (one randomized, replicated, controlled) from the UK found that after specific plants were controlled, new plants established or diversity increased. A replicated, controlled laboratory and grassland study found a specific herbicide had negative impacts on one beetle species. Eleven studies investigated different methods of controlling plants. *Assessment: unknown effectiveness — limited evidence (effectiveness 90%; certainty 28%).*

http://www.conservationevidence.com/actions/123

Protect individual nests of ground-nesting birds

Two randomized, replicated, controlled studies from Sweden found nest exclosures increased measures of ground-nesting bird productivity, however both found bird numbers or adult predation rates were unaffected or negatively affected by exclosures. *Assessment: unknown effectiveness — limited evidence (effectiveness 30%; certainty 13%).*

http://www.conservationevidence.com/actions/108

No evidence found (no assessment)

We have captured no evidence for the following interventions:

- Control grey squirrels

- Erect predator-proof fencing around important breeding sites for waders
- Manage wild deer numbers
- Remove coarse fish

Evidence not assessed

Control bracken

A systematic review found repeated herbicide applications reduced bracken abundance but cutting may be equally effective. A laboratory trial found the same herbicide could inhibit the growth of mosses under certain conditions. *Assessment: this intervention has not been assessed.*

http://www.conservationevidence.com/actions/105

Control invasive non-native plants on farmland (such as Himalayan balsam, Japanese knotweed)

Two randomized, replicated, controlled trials in the Czech Republic found removing all giant hogweed flower heads at peak flowering time reduced seed production. *Assessment: this intervention has not been assessed.*

http://www.conservationevidence.com/actions/104

Control mink

A systematic review found trapping may be an effective method of reducing American mink populations. A study in the UK found mink were successfully eradicated from a large area by systematic trapping. *Assessment: this intervention has not been assessed.*

http://www.conservationevidence.com/actions/107

Provide medicated grit for grouse

A controlled study from the UK found higher red grouse productivity where medicated grit was provided. *Assessment: this intervention has not been assessed.*

http://www.conservationevidence.com/actions/112

4.11 Threat: Education and awareness

Based on the collated evidence, what is the current assessment of the effectiveness of interventions for education and awareness?	
No evidence found (no assessment)	• Provide specialist advice, assistance preparing conservation plans
Evidence not assessed	• Provide training for land managers, farmers and farm advisers

No evidence found (no assessment)

We have captured no evidence for the following intervention:

- Provide specialist advice, assistance preparing conservation plans

Evidence not assessed

Provide training for land managers, farmers and farm advisers

A study from the UK found farmers who were trained in how to implement agri-environment schemes created better quality wildlife habitat over five years. *Assessment: this intervention has not been assessed.*

http://www.conservationevidence.com/actions/113

5. FOREST CONSERVATION

Har'el Agra, Simon Schowanek, Yohay Carmel, Rebecca K. Smith & Gidi Ne'eman

Expert assessors

Rhett Harrison, Consultative Group on International Agricultural Research, Zambia
Keith Kirby, University of Oxford, UK
Gillian Petrokofsky, Biodiversity Institute Oxford, UK
Rebecca K. Smith, University of Cambridge, UK
William J. Sutherland, University of Cambridge, UK
Tom Swinfield, Royal Society for the Protection of Birds, UK

Scope of assessment: for the conservation of forest habitat (not specific species within forests), including tropical forests, temperate forests, woodland, scrubland, shrubland and dry forests.

Assessed: 2016.

Effectiveness measure is the median % score.

Certainty measure is the median % certainty of evidence, determined by the quantity and quality of the evidence in the synopsis.

Harm measure is the median % score for negative side-effects on the forest habitat of concern.

 http://dx.doi.org/10.11647/OBP.0109.05

This book is meant as a guide to the evidence available for different conservation interventions and as a starting point in assessing their effectiveness. The assessments are based on the available evidence for the target habitat for each intervention. The assessment may therefore refer to different habitat to the one(s) you are considering. Before making any decisions about implementing interventions it is vital that you read the more detailed accounts of the evidence in order to assess their relevance for your study species or system.

Full details of the evidence are available at
www.conservationevidence.com

There may also be significant negative side-effects on the target habitats or other species or communities that have not been identified in this assessment.

A lack of evidence means that we have been unable to assess whether or not an intervention is effective or has any harmful impacts.

5.1 Threat: Residential and commercial development

5.1.1 Housing and urban areas

Based on the collated evidence, what is the current assessment of the effectiveness of interventions for residential and commercial development in housing and urban areas?	
No evidence found (no assessment)	• Compensate for woodland removal with compensatory planting • Incorporate existing trees or woods into the landscape of new developments • Provide legal protection of forests from development

No evidence found (no assessment)

We have captured no evidence for the following interventions:

- Compensate for woodland removal with compensatory planting
- Incorporate existing trees or woods into the landscape of new developments
- Provide legal protection of forests from development

5.1.2 Tourism and recreation areas

Based on the collated evidence, what is the current assessment of the effectiveness of interventions for residential and commercial development in tourism and recreation areas?	
No evidence found (no assessment)	• Adopt ecotourism • Create managed paths/signs to contain disturbance • Re-route paths, control access or close paths • Use warning signs to prevent fire

No evidence found (no assessment)

We have captured no evidence for the following interventions:

- Adopt ecotourism
- Create managed paths/signs to contain disturbance
- Re-route paths, control access or close paths
- Use warning signs to prevent fire

5.2 Threat: Agriculture

5.2.1 Livestock farming

Based on the collated evidence, what is the current assessment of the effectiveness of interventions for livestock farming?	
Likely to be beneficial	• Use wire fences within grazing areas to exclude livestock from specific forest sections
Trade-offs between benefit and harms	• Prevent livestock grazing in forests
Unknown effectiveness (limited evidence)	• Reduce the intensity of livestock grazing in forests • Shorten livestock grazing period or control grazing season in forests
No evidence found (no assessment)	• Provide financial incentives not to graze

Likely to be beneficial

Use wire fences within grazing areas to exclude livestock from specific forest sections

Three of four studies, including one replicated, randomized, controlled study in Kenya, Israel, Mexico and Panama found that excluding livestock using wire fences increased the size, density or number of regenerating trees. One study found no effect on tree size and decreased tree density. Four of eight studies, including two replicated, randomized, controlled studies across the world found that excluding livestock using increased biomass, species richness, density or cover of understory plants. Four studies found mixed or no effects on understory plants. *Assessment: likely to be beneficial (effectiveness 58%; certainty 63%; harms 18%).*

http://www.conservationevidence.com/actions/1205

Trade-off between benefit and harms

Prevent livestock grazing in forests

One site comparison study in Israel found that preventing cattle grazing increased the density of seedlings and saplings. Two of three studies, including one replicated, controlled study, in Brazil, Costa Rica and the UK found that preventing livestock grazing increased survival, species richness or diversity of understory plants. One study found mixed effects. *Assessment: trade-offs between benefits and harms (effectiveness 69%; certainty 45%; harms 20%).*

http://www.conservationevidence.com/actions/1206

Unknown effectiveness (limited evidence)

Reduce the intensity of livestock grazing in forests

Two studies, including one replicated, randomized, controlled study, in the UK and Greece found that reducing grazing intensity increased the number of tree saplings or understory total weight. *Assessment: unknown effectiveness (effectiveness 78%; certainty 34%; harms 0%).*

http://www.conservationevidence.com/actions/1207

Shorten livestock grazing period or control grazing season in forests

One of two studies, including one replicated, controlled study, in Spain and Australia found that shortening the grazing period increased the abundance and size of regenerating trees. One found no effect native plant species richness. One replicated study in the UK found that numbers of tree seedlings were higher following summer compared to winter grazing. *Assessment: unknown effectiveness (effectiveness 58%; certainty 33%; harms 0%).*

http://www.conservationevidence.com/actions/1208

No evidence found (no assessment)

We have captured no evidence for the following interventions:

- Provide financial incentives not to graze

5.3 Threat: Transport and service corridors

Based on the collated evidence, what is the current assessment of the effectiveness of interventions for transport and service corridors?	
No evidence found (no assessment)	• Maintain/create habitat corridors

No evidence found (no assessment)

We have captured no evidence for the following interventions:

- Maintain/create habitat corridors

5.4 Threat: Biological resource use

5.4.1 Thinning and wood harvesting

Based on the collated evidence, what is the current assessment of the effectiveness of interventions for thinning and wood harvesting?	
Beneficial	• Log/remove trees within forests: effect on understory plants
Likely to be beneficial	• Thin trees within forests: effects on understory plants • Thin trees within forests: effects on young trees • Use shelterwood harvest instead of clearcutting
Trade-offs between benefit and harms	• Thin trees within forests: effects on mature trees
Unknown effectiveness (limited evidence)	• Log/remove trees within forests: effects on young trees • Use partial retention harvesting instead of clearcutting • Use summer instead of winter harvesting
Unlikely to be beneficial	• Remove woody debris after timber harvest
Likely to be ineffective or harmful	• Log/remove trees within forests: effect on mature trees • Log/remove trees within forests: effect on non-vascular plants • Thin trees within forests: effect on non-vascular plants
No evidence found (no assessment)	• Adopt continuous cover forestry • Use brash mats during harvesting to avoid soil compaction

Beneficial

Log/remove trees within forests: effects on understory plants

Eight of 12 studies, including four replicated, randomized, controlled studies, in India, Australia, Bolivia, Canada and the USA found that logging increased the density and cover or species richness and diversity of understory plants. Two studies found mixed and three found no effect. *Assessment: beneficial (effectiveness 65%; certainty 65%; harms 10%).*

http://www.conservationevidence.com/actions/1273

Likely to be beneficial

Thin trees within forests: effects on understory plants

Twenty five of 38 studies, including 12 replicated, randomized, controlled studies, across the world found that thinning trees increased the density and cover or species richness and diversity of understory plants. Nine studies found mixed and two no effects, and one found a decrease the abundance of herbaceous species. *Assessment: Likely to be beneficial (effectiveness 58%; certainty 73%; harms 13%).*

http://www.conservationevidence.com/actions/1211

Thin trees within forests: effects on young trees

Six of 12 studies, including two replicated, randomized, controlled studies, in Japan and the USA found that thinning increased the density of young trees and a study in Peru found it increased the growth rate of young trees. One study found thinning decreased the density and five found mixed or no effect on young trees. One replicated, controlled study in the USA found no effect on the density of oak acorns. *Assessment: Likely to be beneficial (effectiveness 60%; certainty 65%; harms 15%).*

http://www.conservationevidence.com/actions/1210

Use shelterwood harvest instead of clearcutting

Three replicated, controlled studies in Sweden and the USA found that shelterwood harvesting increased density of trees or plant diversity, or decreased grass cover compared with clearcutting. *Assessment: Likely to be beneficial (effectiveness 75%; certainty 55%; harms 15%).*

http://www.conservationevidence.com/actions/1214

Trade-off between benefit and harms

Thin trees within forests: effects on mature trees

Eleven of 12 studies, including two replicated, randomized, controlled studies, in Brazil, Canada, and the USA found that thinning trees decreased the density and cover of mature trees and in one case tree species diversity. Five of six studies, including one replicated, controlled, before-and-after study, in Australia, Sweden and the USA found that thinning increased mature tree size, the other found mixed effects. One of three studies, including two replicated controlled studies, in the USA found that thinning reduced the number of trees killed by beetles. *Assessment: trade-offs between benefits and harms (effectiveness 47%; certainty 55%; harms 35%).*

http://www.conservationevidence.com/actions/1209

Unknown effectiveness (limited evidence)

Log/remove trees within forests: effects on young trees

One of two replicated controlled studies in Canada and Costa Rica found that logging increased the density of young trees, the other found mixed effects. *Assessment: unknown effectiveness (effectiveness 50%; certainty 18%; harms 10%).*

http://www.conservationevidence.com/actions/1272

Use partial retention harvesting instead of clearcutting

Three studies, including one replicated, randomized, controlled study, in Canada found that using partial retention harvesting instead of clearcutting decreased the density of young trees. *Assessment: unknown effectiveness (effectiveness 5%; certainty 35%; harms 45%).*

http://www.conservationevidence.com/actions/1215

Use summer instead of winter harvesting

One replicated study in the USA found no effect of logging season on plant species richness and diversity. *Assessment: unknown effectiveness (effectiveness 0%; certainty 13%; harms 0%).*

http://www.conservationevidence.com/actions/1216

Unlikely to be beneficial

Remove woody debris after timber harvest

Two studies, including one replicated, randomized, controlled study, in France and the USA found no effect of woody debris removal on cover or species diversity of trees. One of six studies, including two replicated, randomized, controlled studies, in Ethiopia, Spain, Canada and the USA found that woody debris removal increased young tree density. One found that it decreased young tree density and three found mixed or no effect on density or survival. One of six studies, including two replicated, randomized, controlled studies, in the USA and France found that woody debris removal increased understory vegetation cover. Five studies found mixed or no effects on understory vegetation cover or species richness and diversity. *Assessment: unlikely to be beneficial (effectiveness 23%; certainty 50%; harms 10%).*

http://www.conservationevidence.com/actions/1213

Likely to be ineffective or harmful

Log/remove trees within forests: effect on mature trees

Three of seven studies, including two replicated, controlled studies, across the world found that logging trees decreased the density and cover of mature trees. Two found it increased tree density and two found no effect. Four of nine studies, including one replicated, randomized, controlled study, across the world found that logging increased mature tree size or diversity. Four found it decreased tree size or species richness and diversity, and two found no effect on mature tree size or diversity. One replicated, controlled study in Canada found that logging increased mature tree mortality rate. *Assessment: likely to be ineffective or harmful (effectiveness 35%; certainty 50%; harms 30%).*

http://www.conservationevidence.com/actions/1271

Log/remove trees within forests: effect on effects on non-vascular plants

Two of three studies, including one replicated, paired sites study, in Australia, Norway and Sweden found that logging decreased epiphytic plant abundance and fern fertility. One found mixed effects depending on species. *Assessment: likely to be ineffective or harmful (effectiveness 18%; certainty 40%; harms 50%).*

http://www.conservationevidence.com/actions/1270

Thin trees within forests: effects on non-vascular plants

Three of four studies, including one replicated, randomized, controlled study, in Canada, Finland and Sweden found that thinning decreased epiphytic plant abundance and species richness. Three found mixed effects depending on thinning method and species. *Assessment: likely to be ineffective or harmful (effectiveness 20%; certainty 48%; harms 50%).*

http://www.conservationevidence.com/actions/1212

No evidence found (no assessment)

We have captured no evidence for the following interventions:

- Adopt continuous cover forestry
- Use brash mats during harvesting to avoid soil compaction

5.4.2 Harvest forest products

Based on the collated evidence, what is the current assessment of the effectiveness of interventions for harvesting forest products?	
Unknown effectiveness (limited evidence)	• Adopt certification
No evidence found (no assessment)	• Sustainable management of non-timber products

Unknown effectiveness (limited evidence)

Adopt certification

One replicated, site comparison study in Ethiopia found that deforestation risk was lower in certified than uncertified forests. One controlled, before-and-after trial in Gabon found that, when corrected for logging intensity, although tree damage did not differ, changes in above-ground biomass were smaller in certified than in uncertified forests. *Assessment: unknown effectiveness (effectiveness 50%; certainty 20%; harms 3%).*

http://www.conservationevidence.com/actions/1150

No evidence found (no assessment)

We have captured no evidence for the following interventions:

- Sustainable management of non-timber products

5.4.3 Firewood

Based on the collated evidence, what is the current assessment of the effectiveness of interventions for firewood?	
No evidence found (no assessment)	• Provide fuel efficient stoves • Provide paraffin stoves

No evidence found (no assessment)

We have captured no evidence for the following interventions:

- Provide fuel efficient stoves
- Provide paraffin stoves

5.5 Habitat protection

5.5.1 Changing fire frequency

Based on the collated evidence, what is the current assessment of the effectiveness of interventions for changing fire frequency?	
Trade-offs between benefit and harms	• Use prescribed fire: effect on understory plants • Use prescribed fire: effect on young trees
Likely to be ineffective or harmful	• Use prescribed fire: effect on mature trees
No evidence found (no assessment)	• Mechanically remove understory vegetation to reduce wildfires • Use herbicides to remove understory vegetation to reduce wildfires

Trade-off between benefit and harms

Use prescribed fire: effect on understory plants

Eight of 22 studies, including seven replicated, randomized, controlled studies, in Australia, Canada and the USA found that prescribed fire increased the cover, density or biomass of understory plants. Six found it decreased plant cover and eight found mixed or no effect on cover or density. Fourteen of 24 studies, including 10 replicated, randomized, controlled studies, in Australia, France, West Africa and the USA found that fire increased species richness and diversity of understory plants.

One found it decreased species richness and nine found mixed or no effect on understory plants. *Assessment: trade-offs between benefits and harms (effectiveness 55%; certainty 70%; harms 25%).*

http://www.conservationevidence.com/actions/1221

Use prescribed fire: effect on young trees

Five of 15 studies, including four replicated, randomized, controlled studies, in France, Canada and the USA found that prescribed fire increased the density and biomass of young trees. Two found that fire decreased young tree density. Eight found mixed or no effect on density and two found mixed effects on species diversity of young trees. Two replicated, controlled studies in the USA found mixed effects of prescribed fire on young tree survival. *Assessment: trade-offs between benefits and harms (effectiveness 45%; certainty 55%; harms 23%).*

http://www.conservationevidence.com/actions/1220

Likely to be ineffective or harmful

Use prescribed fire: effect on mature trees

Four of nine studies, including two replicated, randomized, controlled studies, in the USA found that prescribed fire decreased mature tree cover, density or diversity. Two studies found it increased tree cover or size, and four found mixed or no effect. Seven studies, including one replicated, randomized, controlled study, in the USA found that fire increased mature tree mortality. *Assessment: likely to be ineffective or harmful (effectiveness 25%; certainty 50%; harms 50%).*

http://www.conservationevidence.com/actions/1217

No evidence found (no assessment)

We have captured no evidence for the following interventions:

- Mechanically remove understory vegetation to reduce wildfires
- Use herbicides to remove understory vegetation to reduce wildfires

5.5.2 Water management

Based on the collated evidence, what is the current assessment of the effectiveness of interventions for water management?	
No evidence found (no assessment)	• Construct water detention areas to slow water flow and restore riparian forests • Introduce beavers to impede water flow in forest watercourses • Recharge groundwater to restore wetland forest

No evidence found (no assessment)

We have captured no evidence for the following interventions:

- Construct water detention areas to slow water flow and restore riparian forests
- Introduce beavers to impede water flow in forest watercourses
- Recharge groundwater to restore wetland forest

5.5.3 Changing disturbance regime

Based on the collated evidence, what is the current assessment of the effectiveness of interventions for changing the disturbance regime?	
Trade-offs between benefit and harms	• Use clearcutting to increase understory diversity • Use group-selection harvesting • Use shelterwood harvesting
Unknown effectiveness (limited evidence)	• Thin trees by girdling (cutting rings around tree trunks) • Use herbicides to thin trees
Unlikely to be beneficial	• Use thinning followed by prescribed fire
No evidence found (no assessment)	• Adopt conservation grazing of woodland • Coppice trees • Halo ancient trees • Imitate natural disturbances by pushing over trees

	• Pollard trees (top cutting or top pruning) • Reintroduce large herbivores • Retain fallen trees

Trade-off between benefit and harms

Use clearcutting to increase understory diversity

Three of nine studies, including four replicated, randomized, controlled studies, in Australia, Japan, Brazil, Canada and the USA found that clearcutting decreased density, species richness or diversity of mature trees. One study found it increased trees species richness and six found mixed or no effect or mixed effect on density, size, species richness or diversity. One replicated, randomized, controlled study in Finland found that clearcutting decreased total forest biomass, particularly of evergreen shrubs. Three of six studies, including five replicated, randomized, controlled studies, in Brazil, Canada and Spain found that clearcutting increased the density and species richness of young trees. One found it decreased young tree density and two found mixed or no effect. Eight of 12 studies, including three replicated, randomized, controlled studies, across the world found that clearcutting increased the cover or species richness of understory plants. Two found it decreased density or species richness, and two found mixed or no effect. *Assessment: trade-offs between benefits and harms (effectiveness 63%; certainty 65%; harms 30%).*

http://www.conservationevidence.com/actions/1222

Use group-selection harvesting

Four of eight studies, including one replicated, controlled study, in Australia, Canada, Costa Rica and the USA found that group-selection harvesting increased cover or diversity of understory plants, or the density of young trees. Two studies found it decreased understory species richness or and biomass. Three studies found no effect on understory species richness or diversity or tree density or growth-rate. *Assessment: trade-offs between benefits and harms (effectiveness 50%; certainty 58%; harms 30%).*

http://www.conservationevidence.com/actions/1224

Use shelterwood harvesting

Six of seven studies, including five replicated, controlled studies, in Australia, Iran, Nepal and the USA found that shelterwood harvesting increased abundance, species richness or diversity or understory plants, as well as the growth and survival rate of young trees. One study found shelterwood harvesting decreased plant species richness and abundance and one found no effect on abundance. One replicated, controlled study in Canada found no effect on oak acorn production. *Assessment: trade-offs between benefits and harms (effectiveness 78%; certainty 70%; harms 28%).*

http://www.conservationevidence.com/actions/1223

Unknown effectiveness (limited evidence)

Thin trees by girdling (cutting rings around tree trunks)

One before-and-after study in Canada found that thinning trees by girdling increased understory plant species richness, diversity and cover. *Assessment: unknown effectiveness — limited evidence (effectiveness 58%; certainty 13%; harms 0%).*

http://www.conservationevidence.com/actions/1226

Use herbicides to thin trees

One replicated, controlled study in Canada found no effect of using herbicide to thin trees on total plant species richness.

Assessment: unknown effectiveness — limited evidence (effectiveness 5%; certainty 13%; harms 0%).

http://www.conservationevidence.com/actions/1225

Unlikely to be beneficial

Use thinning followed by prescribed fire

Three of six studies, including one replicated, randomized, controlled study, in the USA found that thinning followed by prescribed fire increased cover or abundance of understory plants, and density of deciduous trees. One study found it decreased tree density and species richness. Three studies found mixed or no effect or mixed effect on tree growth rate or density of young trees. One replicated, controlled study Australia found no effect of thinning then burning on the genetic diversity of black ash. *Assessment: unlikely to be beneficial (effectiveness 35%; certainty 40%; harms 15%).*

http://www.conservationevidence.com/actions/1227

No evidence found (no assessment)

We have captured no evidence for the following interventions:

- Adopt conservation grazing of woodland
- Coppice trees
- Halo ancient trees
- Imitate natural disturbances by pushing over trees
- Pollard trees (top cutting or top pruning)
- Reintroduce large herbivores
- Retain fallen trees

5.6 Threat: Invasive and other problematic species

5.6.1 Invasive plants

Based on the collated evidence, what is the current assessment of the effectiveness of interventions for invasive plants?	
Unknown effectiveness (limited evidence)	• Manually/mechanically remove invasive plants • Use herbicides to remove invasive plant species
No evidence found (no assessment)	• Use grazing to remove invasive plant species • Use prescribed fire to remove invasive plant species

Unknown effectiveness (limited evidence)

Manually/mechanically remove invasive plants

Two replicated, controlled studies in Hawaii and Ghana found that removing invasive grass or weed species increased understory plant biomass or tree seedling height. Two replicated, controlled studies in the USA and Hawaii found no effect of removing invasive shrubs or plants on understory plant diversity or growth rate of native species. *Assessment: unknown effectiveness — limited evidence (effectiveness 40%; certainty 33%; harms 15%).*

http://www.conservationevidence.com/actions/1228

Use herbicides to remove invasive plant species

One replicated, randomized, controlled study in the USA found no effect of controlling invasive plants using herbicide on native plant species richness. *Assessment: unknown effectiveness — limited evidence (effectiveness 5%; certainty 10%; harms 0%).*

http://www.conservationevidence.com/actions/1229

No evidence found (no assessment)

We have captured no evidence for the following interventions:

- Use grazing to remove invasive plant species
- Use prescribed fire to remove invasive plant species

5.6.2 Native plants

Based on the collated evidence, what is the current assessment of the effectiveness of interventions for native plants?	
No evidence found (no assessment)	• Manually/mechanically remove native plants

No evidence found (no assessment)

We have captured no evidence for the following interventions:

- Manually/mechanically remove native plants

5.6.3 Herbivores

Based on the collated evidence, what is the current assessment of the effectiveness of interventions for herbivores?	
Likely to be beneficial	• Use wire fences to exclude large native herbivores
Unknown effectiveness (limited evidence)	• Use electric fencing to exclude large native herbivores

No evidence found (no assessment)	• Control large herbivore populations • Control medium-sized herbivores • Use fencing to enclose large herbivores (e.g. deer)

Likely to be beneficial

Use wire fences to exclude large native herbivores

Two replicated, controlled studies in the USA found that excluding large herbivores increased tree density. One of three studies, including two replicated, paired-sites, before-and-after studies, in Canada, Bhutan and Ireland found that excluding large herbivores increased the biomass of young trees. One found it decreased the density of young trees and one found mixed effects on species. Five of 10 studies, including two replicated, randomized, controlled studies, across the world found that excluding large herbivores increased the cover or and size of understory plants. Six found no effect on the cover, seed density, species richness or diversity of understory plants. *Assessment: Likely to be beneficial (effectiveness 50%; certainty 65%; harms 10%).*

http://www.conservationevidence.com/actions/1230

Unknown effectiveness (limited evidence)

Use electric fencing to exclude large native herbivores

One controlled study in South Africa found that using electric fencing to exclude elephants and nyalas increased tree density. *Assessment: Unknown effectiveness (effectiveness 65%; certainty 10%; harms 0%).*

http://www.conservationevidence.com/actions/1231

No evidence found (no assessment)

We have captured no evidence for the following interventions:

- Control large herbivore populations
- Control medium-sized herbivores
- Use fencing to enclose large herbivores (e.g. deer)

5.6.4 Rodents

Based on the collated evidence, what is the current assessment of the effectiveness of interventions for rodents?	
Unknown effectiveness (limited evidence)	• Control rodents

Unknown effectiveness (limited evidence)

Control rodents

One controlled study in New Zealand found that rodent control decreased native plant species richness and had no effect on total plant species richness. *Assessment: unknown effectiveness — limited evidence (effectiveness 10%; certainty 10%; harms 50%).*

http://www.conservationevidence.com/actions/1232

5.6.5 Birds

Based on the collated evidence, what is the current assessment of the effectiveness of interventions for birds?	
Unknown effectiveness (limited evidence)	• Control birds

Unknown effectiveness (limited evidence)

Control birds

One controlled study in Australia found that removing birds did not improve the health of the trees in a narrow-leaved peppermint forest. *Assessment: unknown effectiveness — limited evidence (effectiveness 0%; certainty 15%; harms 0%).*

http://www.conservationevidence.com/actions/1151

5.7 Threat: Pollution

Based on the collated evidence, what is the current assessment of the effectiveness of interventions for pollution?	
Unknown effectiveness (limited evidence)	• Maintain/create buffer zones
No evidence found (no assessment)	• Remove nitrogen and phosphorus using harvested products

Unknown effectiveness (limited evidence)

Maintain/create buffer zones

One site comparison study in Australia found that a forest edge protected by a planted buffer strip had higher canopy cover and lower stem density, but similar understory species richness to an unbuffered forest edge. *Assessment: unknown effectiveness — limited evidence (effectiveness 50%; certainty 10%; harms 0%).*

http://www.conservationevidence.com/actions/1168

No evidence found (no assessment)

We have captured no evidence for the following interventions:

- Remove nitrogen and phosphorus using harvested products

5.8 Threat: Climate change and severe weather

Based on the collated evidence, what is the current assessment of the effectiveness of interventions for climate change and severe weather?	
No evidence found (no assessment)	• Prevent damage from strong winds

No evidence found (no assessment)

We have captured no evidence for the following interventions:

- Prevent damage from strong winds

5.9 Habitat protection

Based on the collated evidence, what is the current assessment of the effectiveness of interventions for habitat protection?	
Unknown effectiveness (limited evidence)	• Adopt community-based management to protect forests • Legal protection of forests
No evidence found (no assessment)	• Adopt Protected Species legislation (impact on forest management)

Unknown effectiveness (limited evidence)

Adopt community-based management to protect forests

Two studies, including one replicated, before-and-after, site comparison, in Ethiopia and Nepal found that forest cover increased more in community-managed forests than in forests not managed by local communities. However, one replicated, site comparison study in Colombia found that deforestation rates in community-managed forests did not differ from deforestation rates in unmanaged forests. *Assessment: unknown effectiveness — limited evidence (effectiveness 60%; certainty 35%; harms 0%).*

http://www.conservationevidence.com/actions/1152

Legal protection of forests

Two site comparison studies in Nigeria and Iran found that legal protection of forest increased tree species richness and diversity or the density of young trees. One replicated, paired site study in Mexico found no effect of forest protection on seed density and diversity of trees and shrubs. *Assessment: unknown effectiveness — limited evidence (effectiveness 50%; certainty 20%; harms 0%).*

http://www.conservationevidence.com/actions/1233

No evidence found (no assessment)

We have captured no evidence for the following interventions:

- Adopt Protected Species legislation (impact on forest management)

5.10 Habitat restoration and creation

5.10.1 Restoration after wildfire

Based on the collated evidence, what is the current assessment of the effectiveness of interventions for restoration after wildfire?	
Trade-offs between benefit and harms	• Thin trees after wildfire
Unknown effectiveness (limited evidence)	• Remove burned trees
Likely to be ineffective or harmful	• Sow tree seeds after wildfire
No evidence found (no assessment)	• Plant trees after wildfire

Trade-off between benefit and harms

Thin trees after wildfire

Four of five replicated, controlled studies in Spain, Israel, Cananda and the USA found that thinning trees in burnt forest areas increased plant species richness, cover or survival of saplings. One study found thinning decreased plant biomass. One paired-site study in Canada found that logging after wildfire decreased species richness and diversity of mosses. *Assessment: trade-offs between benefits and harms (effectiveness 50%; certainty 50%; harms 38%).*

http://www.conservationevidence.com/actions/1234

Unknown effectiveness (limited evidence)

Remove burned trees

Two replicated, controlled studies in Israel and Spain found that removing burned trees increased total plant species richness or the cover and species richness of some plant species. *Assessment: unknown effectiveness (effectiveness 60%; certainty 20%; harms 25%).*

http://www.conservationevidence.com/actions/1237

Likely to be ineffective or harmful

Sow tree seeds after wildfire

Three studies, including one replicated, randomized, controlled study, in the USA found that sowing herbaceous plant seeds in burnt forest areas decreased the density of tree seedlings or the number and cover of native species. All three found no effect of seeding on total plant cover or species richness. *Assessment: likely to be ineffective or harmful (effectiveness 0%; certainty 43%; harms 40%).*

http://www.conservationevidence.com/actions/1236

No evidence found (no assessment)

We have captured no evidence for the following interventions:

- Plant trees after wildfire

5.10.2 Restoration after agriculture

Based on the collated evidence, what is the current assessment of the effectiveness of interventions for restoration after agriculture?	
Unknown effectiveness (limited evidence)	• Restore wood pasture (e.g. introduce grazing)

Unknown effectiveness (limited evidence)

Restore wood pasture (e.g. introduce grazing)

One replicated paired study in Sweden found that partial harvesting in abandoned wood pastures increased tree seedling density, survival and growth. *Assessment: unknown effectiveness (effectiveness 65%; certainty 25%; harms 0%).*

http://www.conservationevidence.com/actions/1164

5.10.3 Manipulate habitat to increase planted tree survival during restoration

Based on the collated evidence, what is the current assessment of the effectiveness of interventions for manipulating habitat to increase planted tree survival during restoration?	
Unknown effectiveness (limited evidence)	• Apply herbicides after restoration planting • Cover the ground using techniques other than plastic mats after restoration planting • Cover the ground with plastic mats after restoration planting • Use selective thinning after restoration planting

Unknown effectiveness (limited evidence)

Apply herbicides after restoration planting

One replicated, randomized, controlled study in the USA found that controlling vegetation using herbicides after restoration planting decreased plant species richness and diversity. *Assessment: unknown effectiveness (effectiveness 45%; certainty 25%; harms 40%).*

http://www.conservationevidence.com/actions/1241

Cover the ground using techniques other than plastic mats after restoration planting

One replicated, randomized, controlled study in the USA found that covering the ground with mulch after planting increased total plant cover. *Assessment: unknown effectiveness (effectiveness 30%; certainty 15%; harms 10%).*

http://www.conservationevidence.com/actions/1240

Cover the ground with plastic mats after restoration planting

One replicated study in Canada found that covering the ground with plastic mats after restoration planting decreased the cover of herbecous plants and grasses. *Assessment: unknown effectiveness (effectiveness 40%; certainty 20%; harms 0%).*

http://www.conservationevidence.com/actions/1239

Use selective thinning after restoration planting

One replicated, paired sites study in Canada found that selective thinning after restoration planting conifers increased the abundance of herbaceous species. *Assessment: unknown effectiveness (effectiveness 43%; certainty 18%; harms 0%).*

http://www.conservationevidence.com/actions/1238

5.10.4 Restore forest community

Based on the collated evidence, what is the current assessment of the effectiveness of interventions for restoring a forest community?	
Unknown effectiveness (limited evidence)	• Build bird-perches to enhance natural seed dispersal • Plant a mixture of tree species to enhance diversity • Sow tree seeds • Water plants to preserve dry tropical forest species
No evidence found (no assessment)	• Restore woodland herbaceous plants using transplants and nursery plugs • Use rotational grazing to restore oak savannas

Unknown effectiveness (limited evidence)

Build bird-perches to enhance natural seed dispersal

One replicated, randomized, controlled, before-and-after study in Brazil found that sowing tree seeds increased the density and species richness of new trees. *Assessment: unknown effectiveness (effectiveness 50%; certainty 13%; harms 0%).*

http://www.conservationevidence.com/actions/1245

Plant a mixture of tree species to enhance diversity

One replicated, randomized, controlled study in Brazil found that planting various tree species increased species richness, but had no effect on the density of new trees. One replicated, controlled study in Greece found that planting native tree species increased total plant species richness, diversity and cover. *Assessment: unknown effectiveness (effectiveness 50%; certainty 28%; harms 0%).*

http://www.conservationevidence.com/actions/1243

Sow tree seeds

One replicated, randomized, controlled, before-and-after study in Brazil found that sowing tree seeds increased the density and species richness of new trees. *Assessment: unknown effectiveness (effectiveness 60%; certainty 13%; harms 0%).*

http://www.conservationevidence.com/actions/1244

Water plants to preserve dry tropical forest species

One replicated, controlled study in Hawaii found that watering plants increased the abundance and biomass of forest plants. *Assessment: unknown effectiveness (effectiveness 65%; certainty 18%; harms 0%).*

http://www.conservationevidence.com/actions/1242

No evidence found (no assessment)

We have captured no evidence for the following interventions:

- Restore woodland herbaceous plants using transplants and nursery plugs
- Use rotational grazing to restore oak savannas

5.10.5 Prevent/encourage leaf litter accumulation

Based on the collated evidence, what is the current assessment of the effectiveness of interventions for preventing/encouraging leaf litter accumulation?	
Unknown effectiveness (limited evidence)	• Remove or disturb leaf litter to enhance germination
No evidence found (no assessment)	• Encourage leaf litter development in new planting

Unknown effectiveness (limited evidence)

Remove or disturb leaf litter to enhance germination

One of two replicated, controlled studies in Poland and Costa Rica found that removing leaf litter increased understory plant species richness. The two studies found that removal decreased understory plant cover or the density of new tree seedlings. *Assessment: unknown effectiveness (effectiveness 40%; certainty 25%; harms 23%).*

http://www.conservationevidence.com/actions/1246

No evidence found (no assessment)

We have captured no evidence for the following interventions:

- Encourage leaf litter development in new planting

5.10.6 Increase soil fertility

Based on the collated evidence, what is the current assessment of the effectiveness of interventions for increasing soil fertility?	
Likely to be beneficial	• Use vegetation removal together with mechanical disturbance to the soil
Trade-offs between benefit and harms	• Add organic matter • Use fertilizer • Use soil scarification or ploughing to enhance germination

Unknown effectiveness (limited evidence)	• Add lime to the soil to increase fertility • Use soil disturbance to enhance germination (excluding scarification or ploughing)
Likely to be ineffective or harmful	• Enhance soil compaction

Likely to be beneficial

Use vegetation removal together with mechanical disturbance to the soil

Three studies, including one replicated, randomized, controlled study, in Portugal and France found that vegetation removal together with mechanical disturbance of the soil increased the cover or diversity of understory plants, or density of young trees. One of the studies found it decreased understory shrub cover. *Assessment: Likely to be beneficial (effectiveness 61%; certainty 40%; harms 15%).*

http://www.conservationevidence.com/actions/1274

Trade-off between benefit and harms

Add organic matter

One of two studies, including one replicated, randomized, controlled study, in Brazil and Costa Rica found that adding leaf litter increased species richness of young trees. One found it decreased young tree density in artificial forest gaps and both found no effect on the density of tree regenerations under intact forest canopy. One of two replicated, controlled study in Portugal and the USA found that adding plant material increased total plant cover. One found mixed effects on cover depending on plant group. *Assessment: trade-offs between benefits and harms (effectiveness 45%; certainty 43%; harms 28%).*

http://www.conservationevidence.com/actions/1250

Use fertilizer

Six of eight studies, including five replicated, randomized, controlled, in Europe, Brazil, Australia and the USA found that applying fertilizer increased total plant cover, understory plant biomass, size of young trees, biomass of grasses or cover of artificially seeded plant species. Five of the studies found no effect on plant biomass, cover, seedling abundance, tree growth or tree seedling diversity. *Assessment: trade-offs between benefits and harms (effectiveness 55%; certainty 65%; harms 25%).*

http://www.conservationevidence.com/actions/1248

Use soil scarification or ploughing to enhance germination

Two studies, including one replicated, randomized, controlled study, in Portugal and the USA found that ploughing increased the cover or diversity of understory plants. Two of five studies, including two replicated, randomized, controlled, in Canada, Brazil, Ethiopia and Sweden found that ploughing increased the density of young trees. One found a decrease in density and two found mixed effects depending on tree species. One replicated, before-and-after trial in Finland found that ploughing decreased the cover of plants living on wood surface. One replicated, controlled study in the USA found that ploughing did not decrease the spreading distance and density of invasive grass seedlings. *Assessment: unknown effectiveness (effectiveness 60%; certainty 50%; harms 25%).*

http://www.conservationevidence.com/actions/1251

Unknown effectiveness (limited evidence)

Add lime to the soil to increase fertility

One replicated, randomized controlled study in the USA found that adding lime increased vegetation cover. *Assessment: unknown effectiveness (effectiveness 80%; certainty 18%; harms 0%).*

http://www.conservationevidence.com/actions/1249

Use soil disturbance to enhance germination (excluding scarification or ploughing)

Two replicated, controlled studies in Canada and Finland found that disturbance of the forest floor decreased understory vegetation cover. *Assessment: unknown effectiveness (effectiveness 30%; certainty 35%; harms 40%).*

http://www.conservationevidence.com/actions/1252

Likely to be ineffective or harmful

Enhance soil compaction

Two of three studies, including two replicated, randomized, controlled studies in Canada and the USA found that soil compaction increased understory plant cover and density. Two found it decreased tree regeneration height or density and understory plant species richness. *Assessment: likely to be ineffective or harmful (effectiveness 28%; certainty 40%; harms 45%).*

http://www.conservationevidence.com/actions/1253

5.11 Actions to improve survival and growth rate of planted trees

Based on the collated evidence, what is the current assessment of the effectiveness of interventions to improve the survival and growth rate of planted trees?	
Beneficial	• Prepare the ground before tree planting • Use mechanical thinning before or after planting
Likely to be beneficial	• Fence to prevent grazing after tree planting • Use herbicide after tree planting
Trade-offs between benefit and harms	• Use prescribed fire after tree planting
Unknown effectiveness (limited evidence)	• Apply insecticide to protect seedlings from invertebrates • Add lime to the soil after tree planting • Add organic matter after tree planting • Cover the ground with straw after tree planting • Improve soil quality after tree planting (excluding applying fertilizer) • Manage woody debris before tree planting • Use shading for planted trees • Use tree guards or shelters to protect planted trees • Use weed mats to protect planted trees • Water seedlings
Unlikely to be beneficial	• Mechanically remove understory vegetation after tree planting • Use different planting or seeding methods • Use fertilizer after tree planting

No evidence found (no assessment)	• Apply fungicide to protect seedlings from fungal diseases • Infect tree seedlings with mycorrhizae • Introduce leaf litter to forest stands • Plant a mixture of tree species to enhance the survival and growth of planted trees • Reduce erosion to increase seedling survival • Transplant trees • Use pioneer plants or crops as nurse-plants

Beneficial

Prepare the ground before tree planting

Six of seven studies, including five replicated, randomized, controlled studies, in Canada and Sweden found that ground preparation increased the survival or growth rate of planted trees. One study found no effect of creating mounds on frost damage to seedlings. *Assessment: beneficial (effectiveness 78%; certainty 73%; harms 0%).*

http://www.conservationevidence.com/actions/1263

Use mechanical thinning before or after planting

Five of six studies, including two replicated, randomized, controlled studies, in Brazil, Canada, Finland, France and the USA found that thinning trees after planting increased survival or size of planted trees. One study found mixed effects on survival and size and one found it decreased their density. One replicated study in the USA found that seedling survival rate increased with the size of the thinned area. *Assessment: beneficial (effectiveness 75%; certainty 63%; harms 10%).*

http://www.conservationevidence.com/actions/1261

Likely to be beneficial

Fence to prevent grazing after tree planting

Four of five studies, including two replicated, randomized, controlled studies, in Finland, Australia, Canada and the USA found that using fences to exclude grazing increased the survival, size or cover of planted trees. Two studies found no effect on survival rate and one found mixed effects on planted tree size. *Assessment: Likely to be beneficial (effectiveness 70%; certainty 50%; harms 0%).*

http://www.conservationevidence.com/actions/1254

Use herbicide after tree planting

Two of three studies, including two replicated, randomized, controlled studies, in Sweden and the USA found that using herbicide increased the size of planted trees. One study found no effect. One replicated, randomized, controlled study in Sweden found no effect of using herbicide on frost damage to seedlings. *Assessment: unlikely to be beneficial (effectiveness 58%; certainty 45%; harms 0%).*

http://www.conservationevidence.com/actions/1262

Trade-off between benefit and harms

Use prescribed fire after tree planting

Two of four studies, including one replicated, randomized, controlled study, in Finland, France and the USA found that using prescribed fire after planting increased the survival and sprouting rate of planted trees. One study found fire decreased planted tree size and one found no effect on the size and survival rate. *Assessment: trade-offs between benefits and harms (effectiveness 50%; certainty 43%; harms 20%).*

http://www.conservationevidence.com/actions/1255

Unknown effectiveness (limited evidence)

Apply insecticide to protect seedlings from invertebrates

One randomized, replicated, controlled study in the USA found that applying insecticide increased tree seedling emergence and survival. *Assessment: unknown effectiveness (effectiveness 70%; certainty 13%; harms 0%).*

http://www.conservationevidence.com/actions/1149

Add lime to the soil after tree planting

One of two replicated, randomized, controlled studies in the USA found that adding lime before restoration planting decreased the survival of pine seedlings. One found no effect on seedling growth. *Assessment: unknown effectiveness (effectiveness 0%; certainty 30%; harms 50%).*

http://www.conservationevidence.com/actions/1259

Add organic matter after tree planting

Two replicated, randomized, controlled studies in the USA found that adding organic matter before restoration planting increased seedling biomass, but decreased seedling emergence or survival. *Assessment: unknown effectiveness (effectiveness 20%; certainty 25%; harms 50%).*

http://www.conservationevidence.com/actions/1258

Cover the ground with straw after tree planting

One replicated, randomized, controlled study in the Czech Republic found that covering the ground with straw, but not bark or fleece, increased the growth rate of planted trees and shrubs. *Assessment: unknown effectiveness (effectiveness 75%; certainty 20%; harms 0%).*

http://www.conservationevidence.com/actions/1266

Improve soil quality after tree planting (excluding applying fertilizer)

Two randomized, replicated, controlled studies in Australia found that different soil enhancers had mixed or no effects on tree seedling survival and height, and no effect on diameter or health. *Assessment: unknown effectiveness (effectiveness 25%; certainty 23%; harms 13%).*

http://www.conservationevidence.com/actions/1153

Manage woody debris before tree planting

One replicated, randomized, controlled study in Canada found that removing woody debris increased the survival rate of planted trees. One replicated, controlled study in the USA found mixed effects on the size of planted trees. *Assessment: unknown effectiveness (effectiveness 40%; certainty 25%; harms 13%).*

http://www.conservationevidence.com/actions/1257

Use shading for planted trees

One replicated, controlled study in Panama found that shading increased the survival rate of planted native tree seedlings. *Assessment: unknown effectiveness (effectiveness 85%; certainty 23%; harms 0%).*

http://www.conservationevidence.com/actions/1269

Use tree guards or shelters to protect planted trees

One replicated, randomized, controlled study in the USA found that using light but not dark coloured plastic tree shelters increased the survival rate of planted tree seedlings. One replicated, controlled study in Hong Kong found that tree guards increased tree height after 37 but not 44 months. *Assessment: unknown effectiveness (effectiveness 60%; certainty 28%; harms 20%).*

http://www.conservationevidence.com/actions/1268

Use weed mats to protect planted trees

One replicated, controlled study in Hong Kong found no effect of using weed mats on seedling height. *Assessment: unknown effectiveness (effectiveness 0%; certainty 18%; harms 0%).*

http://www.conservationevidence.com/actions/1267

Water seedlings

One replicated, randomized, controlled study in Spain found that watering seedlings increased or had no effect on seedling emergence and survival, depending on habitat and water availability. *Assessment: unknown effectiveness (effectiveness 45%; certainty 20%; harms 0%).*

http://www.conservationevidence.com/actions/1154

Unlikely to be beneficial

Mechanically remove understory vegetation after tree planting

Four of five studies, including three replicated, randomized, controlled studies in France, Sweden, Panama, Canada and the USA found no effect of controlling understory vegetation on the emergence, survival, growth rate or frost damage of planted seedlings. One found that removing shrubs increased the growth rate and height of planted seedlings, and another that removing competing herbs increased seedling biomass. *Assessment: unlikely to be beneficial (effectiveness 20%; certainty 50%; harms 0%).*

http://www.conservationevidence.com/actions/1256

Use different planting or seeding methods

Four studies, including one replicated, randomized study, in Australia, Brazil, Costa Rica and Mexico found no effect of planting or seeding methods on the size and survival rate of seedlings. One replicated, controlled study in Brazil found that planting early succession pioneer tree species decreased the height of other planted species. *Assessment: unlikely to be beneficial (effectiveness 0%; certainty 43%; harms 13%).*

http://www.conservationevidence.com/actions/1264

Use fertilizer after tree planting

Two of five studies, including two randomized, replicated, controlled studies, in Canada, Australia, France and Portugal found that applying fertilizer after planting increased the size of the planted trees. Three studies found no effect on the size, survival rate or health of planted trees. One randomized, replicated, controlled study in Australia found that soil enhancers including fertilizer had mixed effects on seedling survival and height. *Assessment: unlikely to be beneficial (effectiveness 38%; certainty 45%; harms 3%).*

http://www.conservationevidence.com/actions/1260

No evidence found (no assessment)

We have captured no evidence for the following interventions:

- Apply fungicide to protect seedlings from fungal diseases
- Infect tree seedlings with mycorrhizae
- Introduce leaf litter to forest stands
- Plant a mixture of tree species to enhance the survival and growth of planted trees
- Reduce erosion to increase seedling survival
- Transplant trees
- Use pioneer plants or crops as nurse-plants

5.12 Education and awareness raising

Based on the collated evidence, what is the current assessment of the effectiveness of interventions to improve education and awareness raising?	
No evidence found (no assessment)	• Provide education programmes about forests • Raise awareness amongst the general public through campaigns and public information

No evidence found (no assessment)

We have captured no evidence for the following interventions:

- Provide education programmes about forests
- Raise awareness amongst the general public through campaigns and public information

6. SOME ASPECTS OF CONTROL OF FRESHWATER INVASIVE SPECIES

David Aldridge, Rebecca K. Smith & William J. Sutherland

Expert assessors

David Aldridge, University of Cambridge, UK
Olaf Booy, Animal and Plant Health Agency, UK
Manuel A. Duenas, Centre for Ecology & Hydrology, UK
Alison Dunn, University of Leeds, UK
Robert Francis, King's College London, UK
Belinda Gallardo, Pyrenean Institute of Ecology, Spain
Jonathan Newman, Centre for Ecology & Hydrology, UK
Nancy Ockendon, University of Cambridge, UK
Trevor Renals, Environment Agency, UK
Emmanuelle Sarat, International Union for Conservation of Nature, France
Sonal Varia, The Centre for Agriculture and Bioscience International, UK
Alexandra Zieritz, University of Nottingham, UK

Scope of assessment: for the control of 11 invasive freshwater species.

Assessed: American bullfrog and *Procambarus* spp. crayfish 2015; all other species 2016.

Effectiveness measure is the median % score for effectiveness.

Certainty measure is the median % certainty of evidence for effectiveness, determined by the quantity and quality of the evidence in the synopsis.

Harm measure is the median % score for negative side-effects to non-target native species. This was not assessed for some species in this chapter.

Potential impacts on non-target species should be considered carefully before implementing any control action.

 http://dx.doi.org/10.11647/OBP.0109.06

This book is meant as a guide to the evidence available for different conservation interventions and as a starting point in assessing their effectiveness. The assessments are based on the available evidence for the target group of species for each intervention. The assessment may therefore refer to different species or habitat to the one(s) you are considering. Before making any decisions about implementing interventions it is vital that you read the more detailed accounts of the evidence in order to assess their relevance for your study species or system.

Full details of the evidence are available at
www.conservationevidence.com

There may also be significant negative side-effects on the target groups or other species or communities that have not been identified in this assessment.

A lack of evidence means that we have been unable to assess whether or not an intervention is effective or has any harmful impacts.

6.1 Threat: Invasive plants

6.1.1 Floating pennywort *Hydrocotyle ranunculoides*

Based on the collated evidence, what is the current assessment of the effectiveness of interventions for controlling floating pennywort?	
Beneficial	• Chemical control using herbicides
Likely to be beneficial	• Flame treatment • Physical removal
Unknown effectiveness (limited evidence)	• Combination treatment using herbicides and physical removal
Unlikely to be beneficial	• Biological control using co-evolved, host-specific herbivores • Use of hydrogen peroxide
No evidence found (no assessment)	• Biological control using fungal-based herbicides • Biological control using native herbivores • Environmental control (e.g. shading, reduced flow, reduction of rooting depth, or dredging) • Excavation of banks • Public education • Use of liquid nitrogen

Beneficial

Chemical control using herbicides

A controlled, replicated field study in the UK found that the herbicide 2,4-D amine achieved almost 100% mortality of floating pennywort, compared with the herbicide glyphosate (applied without an adjuvant) which achieved negligible mortality. *Assessment: beneficial (effectiveness 80%; certainty 70%).*

http://www.conservationevidence.com/actions/1127

Likely to be beneficial

Flame treatment

A controlled, replicated study in the Netherlands found that floating pennywort plants were killed by a three second flame treatment with a three second repeat treatment 11 days later. *Assessment: likely to be beneficial (effectiveness 60%; certainty 50%).*

http://www.conservationevidence.com/actions/1131

Physical removal

Two studies, one in Western Australia and one in the UK, found physical removal did not completely eradicate floating pennywort. *Assessment: likely to be beneficial (effectiveness 40%; certainty 40%).*

http://www.conservationevidence.com/actions/1126

Unknown effectiveness (limited evidence)

Combination treatment using herbicides and physical removal

A before-and-after study in Western Australia found that a combination of cutting followed by a glyphosate chemical treatment, removed floating pennywort. *Assessment: unknown effectiveness (effectiveness 70%; certainty 35%).*

http://www.conservationevidence.com/actions/1128

Unlikely to be beneficial

Biological control using co-evolved, host-specific herbivores

A replicated laboratory and field study in South America found that the South American weevil fed on water pennywort but did not reduce the biomass. *Assessment: unlikely to be beneficial (effectiveness 20%; certainty 50%).*

http://www.conservationevidence.com/actions/1123

Use of hydrogen peroxide

A controlled, replicated study in the Netherlands found that hydrogen peroxide sprayed on potted floating pennywort plants at 30% concentration resulted in curling and transparency of the leaves but did not kill the plants. *Assessment: unlikely to be beneficial (effectiveness 10%; certainty 60%).*

http://www.conservationevidence.com/actions/1129

No evidence found (no assessment)

We have captured no evidence for the following interventions:

- Biological control using fungal-based herbicides
- Biological control using native herbivores
- Environmental control (e.g. shading, reduced flow, reduction of rooting depth, or dredging)
- Excavation of banks
- Public education
- Use of liquid nitrogen

6.1.2 Water primrose *Ludwigia spp.*

Based on the collated evidence, what is the current assessment of the effectiveness of interventions for controlling water primrose?	
Likely to be beneficial	• Biological control using co-evolved, host specific herbivores • Chemical control using herbicides • Combination treatment using herbicides and physical removal
Unlikely to be beneficial	• Physical removal
No evidence found (no assessment)	• Biological control using fungal-based herbicides • Biological control using native herbivores • Environmental control (e.g. shading, altered flow, altered rooting depth, or dredging) • Excavation of banks • Public education • Use of a tarpaulin • Use of flame treatment • Use of hydrogen peroxide • Use of liquid nitrogen • Use of mats placed on the bottom of the water body

Likely to be beneficial

Biological control using co-evolved, host specific herbivores

A controlled, replicated study in China, found a flea beetle caused heavy feeding destruction to the prostrate water primrose. A before-and-after study in the USA found that the introduction of flea beetles to a pond significantly reduced the abundance of large-flower primrose-willow. *Assessment: likely to be beneficial (effectiveness 60%; certainty 50%).*

http://www.conservationevidence.com/actions/1135

Chemical control using herbicides

A controlled, replicated laboratory study in the USA found that the herbicide triclopyr TEA applied at concentrations of 0.25% killed 100% of young cultivated water primrose within two months. A before-and-after field study in the UK found that the herbicide glyphosate caused 97% mortality when mixed with a non-oil based sticking agent and 100% mortality when combined with TopFilm. A controlled, replicated, randomized study in Venezuela, found that use of the herbicide halosulfuron-methyl (Sempra) resulted in a significant reduction in water primrose coverage without apparent toxicity to rice plants. *Assessment: likely to be beneficial (effectiveness 80%; certainty 60%).*

http://www.conservationevidence.com/actions/1139

Combination treatment using herbicides and physical removal

A study in the USA found that application of glyphosate and a surface active agent called Cygnet-Plus followed by removal by mechanical means killed 75% of a long-standing population of water primrose. A study in Australia found that a combination of herbicide application, physical removal, and other actions such as promotion of native plants and mulching reduced the cover of Peruvian primrose-willow by 85–90%. *Assessment: likely to be beneficial (effectiveness 70%; certainty 55%).*

http://www.conservationevidence.com/actions/1140

Unlikely to be beneficial

Physical removal

A study in the USA found that hand pulling and raking water primrose failed to reduce its abundance at one site, whereas hand-pulling from the margins of a pond eradicated a smaller population of water primrose at a second site. *Assessment: unlikely to be beneficial (effectiveness 30%; certainty 50%).*

http://www.conservationevidence.com/actions/1138

No evidence found (no assessment)

We have captured no evidence for the following interventions:

- Biological control using fungal-based herbicides
- Biological control using native herbivores
- Environmental control (e.g. shading, reduced flow, reduction of rooting depth, or dredging)
- Excavation of banks
- Public education
- Use of a tarpaulin
- Use of flame treatment
- Use of hydrogen peroxide
- Use of liquid nitrogen
- Use of mats placed on the bottom of the waterbody

6.1.3 Skunk cabbage *Lysichiton americanus*

Based on the collated evidence, what is the current assessment of the effectiveness of interventions for controlling skunk cabbage?	
Likely to be beneficial	• Chemical control using herbicides • Physical removal
No evidence found (no assessment)	• Biological control using co-evolved, host-specific herbivores • Biological control using fungal-based herbicides • Biological control using native herbivores • Combination treatment using herbicides and physical removal • Environmental control (e.g. shading, or promotion of native plants) • Public education • Use of a tarpaulin • Use of flame treatment • Use of hydrogen peroxide • Use of liquid nitrogen

Likely to be beneficial

Chemical control using herbicides

Two studies in the UK found that application of the chemical 2,4-D amine appeared to be successful in eradicating skunk cabbage stands. One of these studies also found glyphosate eradicated skunk cabbage. However, a study in the UK found that glyphosate did not eradicate skunk cabbage, but resulted in only limited reduced growth of plants. *Assessment: likely to be beneficial (effectiveness 60%; certainty 50%).*

http://www.conservationevidence.com/actions/1102

Physical removal

Two studies in Switzerland and the Netherlands, reported effective removal of recently established skunk cabbage plants using physical removal, one reporting removal of the entire stock within five years. A third study in Germany reported that after four years of a twice yearly full removal programme, a large number of plants still needed to be removed each year. *Assessment: likely to be beneficial (effectiveness 65%; certainty 55%).*

http://www.conservationevidence.com/actions/1101

No evidence found (no assessment)

We have captured no evidence for the following interventions:

- Biological control using co-evolved, host-specific herbivores
- Biological control using fungal-based herbicides
- Biological control using native herbivores
- Combination treatment using herbicides and physical removal
- Environmental control (e.g. shading, or promotion of native plants)
- Public education
- Use of a tarpaulin
- Use of flame treatment
- Use of hydrogen peroxide
- Use of liquid nitrogen

6.1.4 New Zealand pigmyweed *Crassula helmsii*

Based on the collated evidence, what is the current assessment of the effectiveness of interventions for controlling *Crassula helmsii*?	
Beneficial	• Chemical control using herbicides • Decontamination to prevent further spread
Likely to be beneficial	• Use lightproof barriers to control plants • Use salt water to kill plants
Unknown effectiveness (limited evidence)	• Use a combination of control measures
Unlikely to be beneficial	• Use dyes to reduce light levels • Use grazing to control plants • Use hot foam to control plants • Use hydrogen peroxide to control plants
No evidence found (no assessment)	• Alter environmental conditions to control plants (e.g. shading by succession, increasing turbidity, re-profiling or dredging) • Biological control using fungal-based herbicides • Biological control using herbivores • Bury plants • Dry out waterbodies • Physical control using manual/mechanical control or dredging • Plant other species to suppress growth • Public education • Surround with wire mesh • Use flame throwers • Use hot water • Use of liquid nitrogen

Beneficial

Chemical control using herbicides

Seven studies in the UK, including one replicated, controlled study, found that applying glyphosate reduced *Crassula helmsii*. Three out of four studies in the UK, including one controlled study, found that applying diquat or diquat alginate reduced or eradicated *C. helmsii*. One small trial found no effect of diquat on *C. helmsii* cover. One replicated, controlled study in the UK found dichlobenil reduced biomass of submerged *C. helmsii* but one small before-and-after study found no effect of dichlobenil on *C. helmsii*. A replicated, controlled study found that treatment with terbutryne partially reduced biomass of submerged *C. helmsii* and that asulam, 2,4-D amine and dalapon reduced emergent *C. helmsii*. *Assessment: beneficial (effectiveness 78%; certainty 75%).*

http://www.conservationevidence.com/actions/1279

Decontamination to prevent further spread

One controlled, replicated container trial in the UK found that submerging *Crassula helmsii* fragments in hot water led to higher mortality than drying out plants or a control. *Assessment: beneficial (effectiveness 80%; certainty 70%).*

http://www.conservationevidence.com/actions/1308

Likely to be beneficial

Use lightproof barriers to control plants

Five before-and-after studies in the UK found that covering with black sheeting or carpet eradicated or severely reduced cover of *Crassula helmsii*. *Assessment: likely to be beneficial (effectiveness 65%; certainty 50%).*

http://www.conservationevidence.com/actions/1294

Use salt water to kill plants

Two replicated, controlled container trials and two before-and-after field trials in the UK found that seawater eradicated *Crassula helmsii*. *Assessment: likely to be beneficial (effectiveness 80%; certainty 45%).*

http://www.conservationevidence.com/actions/1288

Unknown effectiveness (limited evidence)

Use a combination of control methods

One before-and-after study in the UK found that covering *Crassula helmsii* with carpet followed by treatment with glyphosate killed 80% of the plant. *Assessment: unknown effectiveness (effectiveness 75%; certainty 30%).*

http://www.conservationevidence.com/actions/1313

Unlikely to be beneficial

Use dyes to reduce light levels

One replicated, controlled study in the UK found that applying aquatic dye, along with other treatments, did not reduce cover of *Crassula helmsii*. *Assessment: unlikely to be beneficial (effectiveness 0%; certainty 53%).*

http://www.conservationevidence.com/actions/1293

Use grazing to control plants

One of two replicated, controlled studies in the UK found that excluding grazing reduce abundance and coverage of *Crassula helmsii*. The other study found that ungrazed areas had higher coverage of *C. helmsii* than grazed plots. *Assessment: unlikely to be beneficial (effectiveness 23%; certainty 43%).*

http://www.conservationevidence.com/actions/1301

Use hot foam to control plants

One replicated, controlled study in the UK found that treatment with hot foam, along with other treatments, did not control *Crassula helmsii*. A before-and-after study in the UK found that treatment with hot foam partially destroyed *C. helmsii*. *Assessment: unlikely to be beneficial (effectiveness 20%; certainty 50%).*

http://www.conservationevidence.com/actions/1286

Use hydrogen peroxide to control plants

One controlled tank trial in the UK found that hydrogen peroxide did not control *Crassula helmsii*. *Assessment: unlikely to be beneficial (effectiveness 0%; certainty 50%).*

http://www.conservationevidence.com/actions/1281

No evidence found (no assessment)

We have captured no evidence for the following interventions:

- Alter environmental conditions to control plants (e.g. shading by succession, increasing turbidity, re-profiling or dredging)
- Biological control using fungal-based herbicides
- Biological control using herbivores
- Bury plants
- Dry out waterbodies
- Physical control using manual/mechanical control or dredging
- Plant other species to suppress growth
- Public education
- Surround with wire mesh
- Use flame throwers
- Use hot water
- Use of liquid nitrogen

6.2 Threat: Invasive molluscs

6.2.1 Asian clams

Based on the collated evidence, what is the current assessment of the effectiveness of interventions for controlling Asian clams?	
Beneficial	• Add chemicals to the water • Change salinity of the water • Mechanical removal
Likely to be beneficial	• Change temperature of water • Clean equipment • Use of gas-impermeable barriers
Unlikely to be beneficial	• Reduce oxygen in water
No evidence found (no assessment)	• Change pH of water • Drain the invaded waterbody • Exposure to disease-causing organisms • Exposure to parasites • Hand removal • Public awareness and education

Beneficial

● Add chemicals to the water

Two replicated laboratory studies and one controlled, replicated field study found that chlorine, potassium and copper killed Asian clams. Increasing chemical concentration and water temperature killed more clams in less time. One controlled field trial achieved 80% and 100% mortality of Asian clams using encapsulated control agents (SB1000 and SB2000 respectively) in irrigation systems. *Assessment: beneficial (effectiveness 75%; certainty 70%).*

http://www.conservationevidence.com/actions/1118

Change salinity of water

A controlled, replicated laboratory study from the USA found that exposure to saline water killed all Asian clams. *Assessment: beneficial (effectiveness 65%; certainty 68%).*

http://www.conservationevidence.com/actions/1115

Mechanical removal

A controlled before-and-after study from North America found suction dredging of sediment reduced an Asian clam population by 96%, and these effects persisted for a year. A replicated, controlled, before-and-after field trial in Ireland showed that three types of dredges were effective at removing between 74% and >95% of the Asian clam biomass. *Assessment: beneficial (effectiveness 80%; certainty 78%).*

http://www.conservationevidence.com/actions/1120

Likely to be beneficial

Change temperature of water

A controlled laboratory study from the USA found that exposure to water at temperatures of 37°C and 36°C killed all Asian clams within 2 and 4 days, respectively. *Assessment: likely to be beneficial (effectiveness 60%; certainty 55%).*

http://www.conservationevidence.com/actions/1116

Clean equipment

A field study from Portugal found that mechanical removal, followed by regular cleaning and maintenance of industrial pipes at a power plant permanently removed an Asian clam population. A field study from Portugal found that adding a sand filter to a water treatment plant reduced an Asian clam population. *Assessment: likely to be beneficial (effectiveness 75%; certainty 50%).*

http://www.conservationevidence.com/actions/1119

Use of gas-impermeable barriers

One controlled study from North America found that placing gas impermeable fabric barriers on a lake bottom (several small and one large area) reduced populations of Asian clams. *Assessment: likely to be beneficial (effectiveness 78%; certainty 60%).*

http://www.conservationevidence.com/actions/1117

Unlikely to be beneficial

Reduce oxygen in water

A controlled laboratory study from the USA found that Asian clams were not susceptible to low oxygen levels in the water. *Assessment: unlikely to be beneficial (effectiveness 10%; certainty 50%).*

http://www.conservationevidence.com/actions/1113

No evidence found (no assessment)

We have captured no evidence for the following interventions:

- Change pH of water
- Drain the invaded waterbody
- Exposure to disease-causing organisms
- Exposure to parasites
- Hand removal
- Public awareness and education

6.3 Threat: Invasive crustaceans

6.3.1 Ponto-Caspian gammarids

Based on the collated evidence, what is the current assessment of the effectiveness of interventions for controlling Ponto-Caspian gammarids?	
Likely to be beneficial	• Change salinity of the water • Change water temperature • Dewatering (drying out) habitat • Exposure to parasites
Unlikely to be beneficial	• Add chemicals to water • Change water pH • Control movement of gammarids
No evidence found (no assessment)	• Biological control using predatory fish • Cleaning equipment • Exchange ballast water • Exposure to disease-causing organisms

Likely to be beneficial

Change salinity of the water

One of two replicated studies, including one controlled study, in Canada and the UK found that increasing the salinity level of water killed the majority of invasive shrimp within five hours. One found that increased salinity did not kill invasive killer shrimp. *Assessment: likely to be beneficial (effectiveness 40%; certainty 50%).*

http://www.conservationevidence.com/actions/1091

Change water temperature

A controlled laboratory study from the UK found that heating water in excess of 40°C killed invasive killer shrimps. *Assessment: likely to be beneficial (effectiveness 80%; certainty 50%).*

http://www.conservationevidence.com/actions/1092

Dewatering (drying out) habitat

A replicated, controlled laboratory study from Poland found that lowering water levels in sand (dewatering) killed three species of invasive freshwater shrimp, although one species required water content levels of 4% and below before it was killed. *Assessment: likely to be beneficial (effectiveness 60%; certainty 50%).*

http://www.conservationevidence.com/actions/1094

Exposure to parasites

A replicated, controlled experimental study in Canada found that a parasitic mould reduced populations of freshwater invasive shrimp. *Assessment: likely to be beneficial (effectiveness 50%; certainty 50%).*

http://www.conservationevidence.com/actions/1089

Unlikely to be beneficial

Add chemicals to water

A controlled laboratory study from the UK found that four of nine substances added to freshwater killed invasive killer shrimp, but were impractical (iodine solution, acetic acid, Virkon S and sodium hypochlorite). Five substances did not kill invasive killer shrimp (methanol, citric acid, urea, hydrogen peroxide and sucrose). *Assessment: unlikely to be beneficial (effectiveness 35%; certainty 60%).*

http://www.conservationevidence.com/actions/1095

Change water pH

A controlled laboratory study from the UK found that lowering the pH of water did not kill invasive killer shrimp. *Assessment: unlikely to be beneficial (effectiveness 0%; certainty 50%).*

http://www.conservationevidence.com/actions/1093

Control movement of gammarids

Two replicated studies, including one controlled study, in the USA and UK found that movements of invasive freshwater shrimp slowed down or were stopped when shrimp were placed in water that had been exposed to predatory fish or was carbonated. *Assessment: likely to be beneficial (effectiveness 20%; certainty 40%).*

http://www.conservationevidence.com/actions/1088

No evidence found (no assessment)

We have captured no evidence for the following interventions:

- Biological control using predatory fish
- Cleaning equipment
- Exchange ballast water
- Exposure to disease-causing organisms

6.3.2 *Procambarus* spp. crayfish

Based on the collated evidence, what is the current assessment of the effectiveness of interventions for controlling *Procambarus* spp. crayfish?	
Likely to be beneficial	• Add chemicals to the water • Sterilization of males • Trapping and removal • Trapping combined with encouragement of predators
Unknown effectiveness (limited evidence)	• Create barriers
Unlikely to be beneficial	• Encouraging predators
No evidence found (no assessment)	• Draining the waterway • Food source removal • Relocate vulnerable crayfish • Remove the crayfish by electrofishing

Likely to be beneficial

Add chemicals to the water

One replicated study in Italy found that natural pyrethrum at concentrations of 0.05 mg/l and above was effective at killing red swamp crayfish both in the laboratory and in a river, but not in drained burrows. *Assessment: likely to be beneficial (effectiveness 80%; certainty 50%; harms 0%).*

http://www.conservationevidence.com/actions/1036

Sterilization of males

One replicated laboratory study from Italy found that exposing male red swamp crayfish to X-rays reduced the number of offspring they produced. *Assessment: likely to be beneficial (effectiveness 50%; certainty 40%; harms 0%).*

http://www.conservationevidence.com/actions/1032

Trapping and removal

One controlled, replicated study from Italy found that food (tinned meat) was a more effective bait in trapping red swamp crayfish, than using pheromone treatments or no bait (control). Baiting with food increased trapping success compared to trapping without bait. *Assessment: likely to be beneficial (effectiveness 40%; certainty 60%; harms 0%).*

http://www.conservationevidence.com/actions/1029

Trapping combined with encouragement of predators

One before-and-after study in Switzerland and a replicated, paired site study from Italy found that a combination of trapping and predation was more effective at reducing red swamp crayfish populations than predation alone. *Assessment: likely to be beneficial (effectiveness 50%; certainty 50%; harms 0%).*

http://www.conservationevidence.com/actions/1031

Unknown effectiveness (limited evidence)

Create barriers

One before-and-after study from Italy found that the use of concrete dams across a stream was effective at containing spread of the population upstream. *Assessment: unknown effectiveness (effectiveness 30%; certainty 30%; harms 0%).*

http://www.conservationevidence.com/actions/1037

Unlikely to be beneficial

Encouraging predators

Two replicated, controlled studies in Italy found that eels fed on the red swamp crayfish and reduced population size. One replicated, controlled study found that pike predated red swamp crayfish. *Assessment: unlikely to be beneficial (effectiveness 30%; certainty 60%; harms 0%).*

http://www.conservationevidence.com/actions/1030

No evidence found (no assessment)

We have captured no evidence for the following interventions:

- Draining the waterway
- Food source removal
- Relocate vulnerable crayfish
- Remove the crayfish by electrofishing

6.4 Threat: Invasive fish

6.4.1 Brown and black bullheads

Based on the collated evidence, what is the current assessment of the effectiveness of interventions for controlling brown and black bullheads?	
Beneficial	• Application of a biocide
Likely to be beneficial	• Netting
No evidence found (no assessment)	• Biological control of beneficial species • Biological control using native predators • Changing salinity • Changing pH • Draining invaded waterbodies • Electrofishing • Habitat manipulation • Increasing carbon dioxide concentrations • Public education • Trapping using sound or pheromonal lures • Using a combination of netting and electrofishing • UV radiation

Beneficial

Application of a biocide

Two studies in the UK and USA found that rotenone successfully eradicated black bullhead. *Assessment: beneficial (effectiveness 80%; certainty 80%).*

http://www.conservationevidence.com/actions/1050

Likely to be beneficial

Netting

A replicated study in a nature reserve in Belgium found that double fyke nets could be used to significantly reduce the population of large brown bullheads. *Assessment: likely to be beneficial (effectiveness 55%; certainty 55%).*

http://www.conservationevidence.com/actions/1051

No evidence found (no assessment)

We have captured no evidence for the following interventions:

- Biological control of beneficial species
- Biological control using native predators
- Changing salinity
- Changing pH
- Draining invaded waterbodies
- Electrofishing
- Habitat manipulation
- Increasing carbon dioxide concentrations
- Public education
- Trapping using sound or pheromonal lures
- Using a combination of netting and electrofishing
- UV radiation

6.4.2 Ponto-Caspian gobies

Based on the collated evidence, what is the current assessment of the effectiveness of interventions for controlling Ponto-Caspian gobies?	
Beneficial	• Changing salinity
Likely to be beneficial	• Use of barriers to prevent migration
No evidence found (no assessment)	• Application of a biocide • Biological control of beneficial species • Biological control using native predators • Changing pH • Draining invaded waterbodies • Electrofishing • Habitat manipulation • Increasing carbon dioxide concentrations • Netting • Public education • Trapping using visual, sound and pheromonal lures • Using a combination of netting and electrofishing • UV radiation

Beneficial

● Changing salinity

A replicated controlled laboratory study in Canada found 100% mortality of round gobies within 48 hours of exposure to water of 30% salinity. *Assessment: beneficial (effectiveness 90%; certainty 75%).*

http://www.conservationevidence.com/actions/1072

Likely to be beneficial

Use of barriers to prevent migration

A controlled, replicated field study in the USA found that an electrical barrier prevented movement of round gobies across it, and that increasing electrical pulse duration and voltage increased the effectiveness of the barrier. *Assessment: likely to be beneficial (effectiveness 50%; certainty 45%).*

http://www.conservationevidence.com/actions/1074

No evidence found (no assessment)

We have captured no evidence for the following interventions:

- Application of a biocide
- Biological control of beneficial species
- Biological control using native predators
- Changing pH
- Draining invaded waterbodies
- Electrofishing
- Habitat manipulation
- Increasing carbon dioxide concentrations
- Netting
- Public education
- Trapping using visual, sound and pheromonal lures
- Using a combination of netting and electrofishing
- UV radiation

6.5 Threat: Invasive reptiles

6.5.1 Red-eared terrapin *Trachemys scripta*

Based on the collated evidence, what is the current assessment of the effectiveness of interventions for controlling red-eared terrapin?	
Likely to be beneficial	• Direct removal of adults
Unlikely to be beneficial	• Application of a biocide
No evidence found (no assessment)	• Biological control using native predators • Draining invaded waterbodies • Public education • Search and removal using sniffer dogs

Likely to be beneficial

Direct removal of adults

Two studies, a replicated study from Spain using Aranzadi turtle traps, and an un-replicated study in the British Virgin Islands using sein netting, successfully captured but did not eradicate red-eared terrapin populations. *Assessment: likely to be beneficial (effectiveness 40%; certainty 50%).*

http://www.conservationevidence.com/actions/1055

Unlikely to be beneficial

Application of a biocide

A replicated, controlled laboratory study in the USA, found that application of glyphosate to the eggs of red-eared terrapins reduced hatching success to 73% but only at the highest experimental concentration of glyphosate and a surface active agent. *Assessment: unlikely to be beneficial (effectiveness 15%; certainty 50%).*

http://www.conservationevidence.com/actions/1059

No evidence found (no assessment)

We have captured no evidence for the following interventions:

- Biological control using native predators
- Draining invaded waterbodies
- Public education
- Search and removal using sniffer dogs

6.6 Threat: Invasive amphibians

6.6.1 American bullfrog *Lithobates catesbeiana*

Based on the collated evidence, what is the current assessment of the effectiveness of interventions for controlling American bullfrogs?	
Likely to be beneficial	• Biological control using native predators • Direct removal of adults • Direct removal of juveniles
Unknown effectiveness (limited evidence)	• Application of a biocide
No evidence found (no assessment)	• Biological control of co-occurring beneficial species • Collection of egg clutches • Draining ponds • Fencing • Habitat modification • Pond destruction • Public education

Likely to be beneficial

Biological control using native predators

One replicated, controlled study conducted in northeast Belgium found the introduction of the northern pike led to a strong decline in bullfrog tadpole numbers. *Assessment: likely to be beneficial (effectiveness 70%; certainty 40%; harms 0%).*

http://www.conservationevidence.com/actions/1039

Direct removal of adults

One replicated study in Belgium found catchability of adult bullfrogs in small shallow ponds using a double fyke net to be very low. One small study in the USA found that adult bullfrogs can be captured overnight in a single trap floating on the water surface. One replicated, controlled study in the USA found that bullfrog populations rapidly rebounded following intensive removal of the adults. One study in France found a significant reduction in the number of recorded adults and juveniles following the shooting of metamorphosed individuals before reproduction, when carried out as part of a combination treatment. *Assessment: likely to be beneficial (effectiveness 50%; certainty 70%; harms 0%).*

http://www.conservationevidence.com/actions/1045

Direct removal of juveniles

One replicated study in Belgium found double fyke nets were effective in catching bullfrog tadpoles in small shallow ponds. One study in France found a significant reduction in the number of recorded adults and juveniles following the removal of juveniles by trapping, when carried out as part of a combination treatment. *Assessment: likely to be beneficial (effectiveness 70%; certainty 60%; harms 0%).*

http://www.conservationevidence.com/actions/1046

Unknown effectiveness (limited evidence)

Application of a biocide

One replicated, controlled study in the USA reported a number of chemicals killed American bullfrogs, including caffeine (10% solution), chloroxylenol (5% solution), and a combined treatment of Permethrin (4.6% solution) and Rotenone (1% solution). *Assessment: unknown effectiveness (effectiveness 50%; certainty 20%; harms 0%).*

http://www.conservationevidence.com/actions/1048

No evidence found (no assessment)

We have captured no evidence for the following interventions:

- Biological control of co-occurring beneficial species
- Collection of egg clutches
- Draining ponds
- Fencing
- Habitat modification
- Pond destruction
- Public education

7. SOME ASPECTS OF ENHANCING NATURAL PEST CONTROL

Hugh, L. Wright, Joscelyne E. Ashpole, Lynn V. Dicks, James Hutchison, Caitlin G. McCormack & William J. Sutherland

Expert assessors

Barbara Smith, Game and Wildlife Conservation Trust, UK
Tony Harding, Rothamsted Research, UK
Anthony Goggin, Linking Environment and Farming (LEAF), UK
Felix Wackers, BioBest/University of Lancaster, Belgium/UK
Melvyn Fidgett, Syngenta, UK
Michael Garratt, University of Reading, UK
Michelle Fountain, East Malling Research, UK
Phillip Effingham, Greentech Consultants, UK
Stephanie Williamson, Pesticides Action Network, UK
Toby Bruce, Rothamsted Research, UK
Andrew Wilby, University of Lancaster, UK
Eve Veromann, Estonian University of Life Sciences, Estonia
Mattias Jonsson, Swedish University of Agricultural Sciences, Sweden
Vicky Kindemba, Buglife, UK
Steve Sait, University of Leeds, UK

Scope of assessment: 22 of 92 possible actions to enhance natural regulation of pests (including animals, plants, fungi, bacteria and viruses) in agricultural systems across the world.

Assessed: 2014.

Effectiveness measure is the median % score.

Certainty measure is the median % certainty of evidence, determined by the quantity and quality of the evidence in the synopsis.

Harm measure is the median % score for negative side-effects for the farmer such as reduced yield and profits or increased costs.

 http://dx.doi.org/10.11647/OBP.0109.07

This book is meant as a guide to the evidence available for different conservation interventions and as a starting point in assessing their effectiveness. The assessments are based on the available evidence for the target group of species for each intervention. The assessment may therefore refer to different species or habitat to the one(s) you are considering. Before making any decisions about implementing interventions it is vital that you read the more detailed accounts of the evidence in order to assess their relevance for your study species or system.

Full details of the evidence are available at
www.conservationevidence.com

There may also be significant negative side-effects on the target groups or other species or communities that have not been identified in this assessment.

A lack of evidence means that we have been unable to assess whether or not an intervention is effective or has any harmful impacts.

7.1 Reducing agricultural pollution

Based on the collated evidence, what is the current assessment of the effectiveness of interventions that reduce agricultural pollution for enhancing natural pest regulation?	
Unknown effectiveness (limited evidence)	• Alter the timing of insecticide use • Delay herbicide use • Incorporate parasitism rates when setting thresholds for insecticide use • Use pesticides only when pests or crop damage reach threshold levels
Evidence not assessed	• Convert to organic farming

Unknown effectiveness (limited evidence)

Alter the timing of insecticide use

- *Natural enemies*: One controlled study from the UK reported more natural enemies when insecticides were sprayed earlier rather than later in the growing season.
- *Pests*: Two of four studies from Mozambique, the UK and the USA found fewer pests or less disease damage when insecticides were applied early rather than late. Effects on a disease-carrying pest varied with insecticide type. Two studies (including one randomized, replicated, controlled test) found no effect on pests or pest damage.

- *Yield*: Four studies (including one randomized, replicated, controlled test) from Mozambique, the Philippines, the UK and the USA measured yields. Two studies found mixed effects and one study found no effect on yield when insecticides were applied early. One study found higher yields when insecticides were applied at times of suspected crop susceptibility.
- *Profit and costs:* One controlled study from the Philippines found higher profits and similar costs when insecticides were only applied at times of suspected crop susceptibility.
- *Crops studied:* aubergine, barley, maize, pear, stringbean.
- *Assessment: unknown effectiveness (effectiveness 40%; certainty 28%; harms 13%).*

http://www.conservationevidence.com/actions/723

Delay herbicide use

- *Natural enemies:* Two randomized, replicated, controlled trials from Australia and Denmark found more natural enemies when herbicide treatments were delayed. One of the studies found some but not all natural enemy groups benefited and fewer groups benefitted early in the season.
- *Weeds:* One randomized, replicated, controlled study found more weeds when herbicide treatments were delayed.
- *Insect pests and damage:* One of two randomized, replicated, controlled studies from Canada and Denmark found more insect pests, but only for some pest groups, and one study found fewer pests in one of two experiments and for one of two crop varieties. One study found lower crop damage in some but not all varieties and study years.
- *Yield:* One randomized, replicated, controlled study found lower yields.
- *Crops studied:* beet and oilseed.
- *Assessment: unknown effectiveness (effectiveness 20%; certainty 25%; harms 50%).*

http://www.conservationevidence.com/actions/774

Incorporate parasitism rates when setting thresholds for insecticide use

- *Pest damage:* One controlled study from New Zealand found using parasitism rates to inform spraying decisions resulted in acceptable levels of crop damage from pests. Effects on natural enemy populations were not monitored.
- *The crop studied* was tomato.
- *Assessment: unknown effectiveness (effectiveness 50%; certainty 10%; harms 5%).*

http://www.conservationevidence.com/actions/726

Use pesticides only when pests or crop damage reach threshold levels

- *Natural enemies*: One randomized, replicated, controlled study from Finland found that threshold-based spraying regimes increased numbers of natural enemies in two of three years but effects lasted for as little as three weeks.
- *Pests and disease*: Two of four studies from France, Malaysia and the USA reported that pests were satisfactorily controlled. One randomized, replicated, controlled study found pest numbers were similar under threshold-based and conventional spraying regimes and one study reported that pest control was inadequate. A randomized, replicated, controlled study found mixed effects on disease severity.
- *Crop damage*: Four of five randomized, replicated, controlled studies from New Zealand, the Philippines and the USA found similar crop damage under threshold-based and conventional, preventative spraying regimes, but one study found damage increased. Another study found slightly less crop damage compared to unsprayed controls.
- *Yield:* Two of four randomized, replicated, controlled studies found similar yields under threshold-based and conventional spraying regimes. Two studies found mixed effects depending on site, year, pest stage/type or control treatment.

- *Profit:* Two of three randomized, replicated, controlled studies found similar profits using threshold-based and conventional spraying regimes. One study found effects varied between sites and years.
- *Costs:* Nine studies found fewer pesticide applications were needed and three studies found or predicted lower production costs.
- *Crops studied:* barley, broccoli, cabbages, cauliflower, celery, cocoa, cotton, grape, peanut, potato, rice, tomato, and wheat.
- *Assessment: unknown effectiveness (effectiveness 39%; certainty 30%; harms 20%).*

http://www.conservationevidence.com/actions/750

Evidence not assessed

Convert to organic farming

- *Parasitism and mortality (caused by natural enemies)*: One of five studies (three replicated, controlled tests and two also randomized) from Europe, North America, Asia and Australasia found that organic farming increased parasitism or natural enemy-induced mortality of pests. Two studies found mixed effects of organic farming and two randomized, replicated, controlled studies found no effect.
- *Natural enemies:* Eight of 12 studies (including six randomized, replicated, controlled tests) from Europe, North America Asia and Australasia found more natural enemies under organic farming, although seven of these found effects varied over time or between natural enemy species or groups and/or crops or management practices. Three studies (one randomized, replicated, controlled) found no or inconsistent effects on natural enemies and one study found a negative effect.
- *Pests and diseases:* One of eight studies (including five randomized, replicated, controlled tests) found that organic farming reduced pests or disease, but two studies found more pests. Three studies found mixed effects and two studies found no effect.
- *Crop damage:* One of seven studies (including five randomized, replicated, controlled tests) found less crop damage in organic fields but two studies found more. One study found a mixed response and three studies found no or inconsistent effects.

- *Weed seed predation and weed abundance:* One randomized, replicated, controlled study from the USA found mixed effects of organic farming on weed seed predation by natural enemies. Two of three randomized, replicated, controlled studies from the USA found more weeds in organically farmed fields, but in one of these studies this effect varied between crops and years. One study found no effect.
- *Yield and profit:* Six randomized, replicated, controlled studies measured yields and found one positive effect, one negative effect and one mixed effect, plus no or inconsistent effects in three studies. One study found net profit increased if produce received a premium, but otherwise profit decreased. Another study found a negative or no effect on profit.
- *Crops studied:* apple, barley, beans, cabbage, carrot, gourd, maize, mixed vegetables, pea, pepper, safflower, soybean, tomato and wheat.

http://www.conservationevidence.com/actions/717

7.2 All farming systems

Based on the collated evidence, what is the current assessment of the effectiveness of interventions on all farming systems for enhancing natural pest regulation?	
Likely to be beneficial	• Grow non-crop plants that produce chemicals that attract natural enemies • Use chemicals to attract natural enemies
Trade-offs between benefit and harms	• Leave part of the crop or pasture unharvested or uncut
Unknown effectiveness (limited evidence)	• Plant new hedges • Use alley cropping
Evidence not assessed	• Use mass-emergence devices to increase natural enemy populations

Likely to be beneficial

Grow non-crop plants that produce chemicals that attract natural enemies

- *Natural enemies:* Four studies from China, Germany, India and Kenya tested the effects of growing plants that produce chemicals that attract natural enemies. Three (including one replicated, randomized, controlled trail) found higher numbers of natural enemies in plots with plants that produce attractive chemicals, and one also found that the plant used attracted natural enemies in lab studies. One found no effect on parasitism but the plant used was found not to be attractive to natural enemies in lab studies.

- *Pests:* All four studies found a decrease in either pest population or pest damage in plots with plants that produce chemicals that attract natural enemies.
- *Yield:* One replicated, randomized, controlled study found an increase in crop yield in plots with plants that produce attractive chemicals.
- *Crops studied:* sorghum, safflower, orange and lettuce.
- *Assessment: likely to be beneficial (effectiveness 68%; certainty 40%; harms 0%).*

http://www.conservationevidence.com/actions/724

Use chemicals to attract natural enemies

- *Parasitism and predation (by natural enemies)***:** One review and two of five studies from Asia, Europe and North America found that attractive chemicals increased parasitism. Two studies, including one randomized, replicated, controlled trial, found greater parasitism for some but not all chemicals, crops, sites or years and one study found no effect. One study showed that parasites found pests more rapidly. One study found lower egg predation by natural predators.
- *Natural enemies:* Five of 13 studies from Africa, Asia, Australasia, Europe and North America found more natural enemies while eight (including seven randomized, replicated, controlled trials) found positive effects varied between enemy groups, sites or study dates. Four of 13 studies (including a meta-analysis) found more natural enemies with some but not all test chemicals. Two of four studies (including a review) found higher chemical doses attracted more enemies, but one study found lower doses were more effective and one found no effect.
- *Pests:* Three of nine studies (seven randomized, replicated, controlled) from Asia, Australasia, Europe and North America found fewer pests, although the effect occurred only in the egg stage in one study. Two studies found more pests and four found no effect.
- *Crop damage:* One study found reduced damage with some chemicals but not others, and one study found no effect.
- *Yield:* One study found higher wheat yields.

- *Crops studied:* apple, banana, bean, broccoli, Chinese cabbage, cotton, cowpea, cranberry, grape, grapefruit, hop, maize, oilseed, orange, tomato, turnip and wheat.
- *Assessment: likely to be beneficial (effectiveness 40%; certainty 50%; harms 15%).*

http://www.conservationevidence.com/actions/754

Trade-off between benefit and harms

Leave part of the crop or pasture unharvested or uncut

- *Natural enemies:* We found eight studies from Australia, Germany, Hungary, New Zealand, Switzerland and the USA that tested leaving part of the crop or pasture unharvested or unmown. Three (including one replicated, controlled trial) found an increase in abundance of predatory insects or spiders in the crop field or pasture that was partly uncut, while four (including three replicated, controlled trials), found more predators in the unharvested or unmown area itself. Two studies (one replicated and controlled) found that the ratio of predators to pests was higher in partially cut plots and one replicated, controlled study found the same result in the uncut area. Two replicated, controlled studies found differing effects between species or groups of natural enemies.
- *Predation and parasitism:* One replicated, controlled study from Australia found an increase in predation and parasitism rates of pest eggs in unharvested strips.
- *Pests:* Two studies (including one replicated, controlled study) found a decrease in pest numbers in partially cut plots, one of them only for one species out of two. Two studies (one replicated, the other controlled) found an increase in pest numbers in partially cut plots, and two studies (including one replicated, controlled study) found more pests in uncut areas.
- *Crops studied:* alfalfa and meadow pastures.
- *Assessment: trade-offs between benefits and harms (effectiveness 45%; certainty 50%; harms 25%).*

http://www.conservationevidence.com/actions/725

Unknown effectiveness (limited evidence)

Plant new hedges

- *Natural enemies:* One randomized, replicated, controlled study from China compared plots with and without hedges and found no effect on spiders in crops. One of two studies from France and China found more natural enemies in a hedge than in adjacent crops while one study found this effect varied between crop types, hedge species and years. Two randomized, replicated, controlled studies from France and Kenya found natural enemy abundance in hedges was affected by the type of hedge shrub/ tree planted and one also found this effect varied between natural enemy groups.
- *Pests:* One randomized, replicated, controlled study from Kenya compared fallow plots with and without hedges and found effects varied between nematode (roundworm) groups.
- *Crops studied:* barley, beans, maize and wheat.
- *Assessment: unknown effectiveness (effectiveness 20%; certainty 19%; harms 20%).*

http://www.conservationevidence.com/actions/752

Use alley cropping

- *Parasitism, infection and predation:* Two of four studies from Kenya and the USA (including three randomized, replicated, controlled trials) found that effects of alley cropping on parasitism varied between study sites, sampling dates, pest life stages or the width of crop alleys. Two studies found no effect on parasitism. One study found mixed effects on fungal infections in pests and one study found lower egg predation.
- *Natural enemies:* One randomized, replicated, controlled study from Kenya found more wasps and spiders but fewer ladybirds. Some natural enemy groups were affected by the types of trees used in hedges.
- *Pests and crop damage:* Two of four replicated, controlled studies (two also randomized) from Kenya, the Philippines and the UK found more pests in alley cropped plots. One study found fewer pests and

one study found effects varied with pest group and between years. One study found more pest damage to crops but another study found no effect.

- *Weeds:* One randomized, replicated, controlled study from the Philippines found mixed effects on weeds, with more grasses in alley cropped than conventional fields under some soil conditions.
- *Yield:* One controlled study from the USA found lower yield and one study from the Philippines reported similar or lower yields.
- *Costs and profit:* One study from the USA found lower costs but also lower profit in alley cropped plots.
- *Crops studied:* alfalfa, barley, cowpea, maize, pea, rice and wheat.
- *Assessment: unknown effectiveness (effectiveness 15%; certainty 35%; harms 50%).*

http://www.conservationevidence.com/actions/718

Evidence not assessed

Use mass-emergence devices to increase natural enemy populations

- *Parasitism:* One randomized, replicated, controlled study in Switzerland found higher parasitism at one site but no effect at another site when mass-emergence devices were used in urban areas.
- *Pest damage:* The same study found no effect on pest damage to horse chestnut trees.

http://www.conservationevidence.com/actions/775

7.3 Arable farming

Based on the collated evidence, what is the current assessment of the effectiveness of interventions on arable farming systems for enhancing natural pest regulation?	
Beneficial	• Combine trap and repellent crops in a push-pull system
Trade-offs between benefit and harms	• Use crop rotation in potato farming systems
Unlikely to be beneficial	• Create beetle banks
Likely to be ineffective or harmful	• Incorporate plant remains into the soil that produce weed-controlling chemicals

Beneficial

Combine trap and repellent crops in a push-pull system

- *Parasitism:* Two randomized, replicated, controlled studies from Kenya found that push-pull cropping systems increased parasitism of stem borer larvae. One of the studies found no effect on egg parasitism.
- *Natural enemies:* Two randomized, replicated, controlled studies from Kenya and South Africa found push-pull systems had more natural predators, both in overall totals and the abundance of different predator groups.
- *Pests:* Two of three studies (two randomized, replicated, controlled) in Ethiopia, Kenya and South Africa found fewer pests. One study

found no effect on pest infestation, but pests were scarce throughout. Two replicated, controlled studies (one also randomized) found fewer witchweeds.

- *Crop damage*: Two of three replicated, controlled studies (one randomized) found less pest damage, but one study (where pest numbers were low) found effects varied between years and types of damage symptom.
- *Yield:* Four of five replicated, controlled studies (two also randomized) found higher yields and one found no effect.
- *Profit and cost:* Two studies in Kenya and a review found greater economic benefits. One study found higher production costs in the first year, but equal or lower costs in the following five years.
- *Crops studied:* maize and beans.
- *Assessment: beneficial (effectiveness 70%; certainty 68%; harms 5%).*

http://www.conservationevidence.com/actions/753

Trade-off between benefit and harms

Use crop rotation in potato farming systems

- *Pests:* Nine studies from Canada and the USA and one review investigated the effect of crop rotation on pest or pathogen populations in potato. Three studies (including two replicated studies of which one randomized and one controlled) and a review found crop rotation reduced pest populations and crop diseases in at least one year or at least one site. One paired study found pest populations increased in crop rotation. Four studies (including one replicated, randomized, controlled trial) found increases and decreases in pest populations depending on rotation crops used and other treatments. One replicated, randomized, controlled study6 found no effect.
- *Yield:* Three out of five studies (all replicated, controlled, two also randomized) from Canada and the USA, found that crop rotation increased crop yield in some years or with certain rotation crops. The two other studies (both replicated, one also randomized and one replicated) found yield increases and decreases depending on rotation crops used.

- *Profit:* One replicated, controlled study found that crop rotation increased profit.
- *Insecticides:* Two studies (one replicated, controlled) found that fewer insecticide treatments were needed on rotated plots.
- *Crops studied:* alfalfa, barley, broccoli, brown mustard, buckwheat, cotton, lupins, maize, oats, pearl millet, peas, potato, rye, sorghum, soybean, sugar beet, timothy grass, wheat and yellow sweet clover.
- *Assessment: trade-offs between benefits and harms (effectiveness 50%; certainty 50%; harms 25%).*

http://www.conservationevidence.com/actions/719

Unlikely to be beneficial

Create beetle banks

- *Natural enemies in fields:* Six studies from Canada, the UK and USA (three replicated, controlled, of which two were also randomized) examined the effects on predator numbers in adjacent crops. A review found that predators increased in adjacent crops, but one study found effects varied with time and another found no effect. Two studies found small or slow movements of predators from banks to crops. One study found greater beetle activity in fields but this did not improve pest predation.
- *Natural enemies on banks:* Four studies and a review found more invertebrate predators on beetle banks than in surrounding crops, but one of these found that effects varied with time.
- Eight studies from the UK and USA (including two randomized, replicated, controlled trials and two reviews) compared numbers of predatory invertebrates on beetle banks with other refuge habitats. Two studies found more natural enemies on beetle banks, but one of these found only seasonal effects. One review found similar or higher numbers of predators on beetle banks and four studies found similar or lower numbers.
- *Pests:* A replicated, randomized, controlled study and a review found the largest pest reductions in areas closest to a beetle bank or on the beetle bank itself. One review found fewer pests in fields with than without a beetle bank.

- *Economics:* One replicated, randomized, controlled trial and a review showed that beetle banks could make economic savings if they prevented pests from reaching a spray threshold or causing 5% yield loss.
- *Beetle bank design:* Two studies from the UK found certain grass species held higher numbers of predatory invertebrates than others.
- *Crops studied:* barley, field bean, maize, oats, pasture, pea, radish, rapeseed, soybean and wheat.
- *Assessment: unlikely to be beneficial (effectiveness 25%; certainty 60%; harms 10%).*

http://www.conservationevidence.com/actions/729

Likely to be ineffective or harmful

Incorporate plant remains into the soil that produce weed-controlling chemicals

- *Weeds:* Six studies (including six randomized, replicated, controlled tests) from Asia, Europe and North America examined the effect of allelopathic plant residues on weeds by comparing amended soils with weeded controls. Three studies found a reduction in weed growth, and three found effects varied between years, weed groups, or type of weeding method in controls.
- Four studies from Asia and North America examined the effect on weeds by comparing amended soils with unweeded controls. Two studies found a reduction in weed growth, but one found that residues applied too far in advance of crop planting had the reverse effect.
- Two studies found that effects varied between trials, weed species or the type of residue used.
- *Weed control:* Two studies, including one randomized, replicated, controlled laboratory study, found that the decrease in weeds did not last beyond a few days or weeks after residue incorporation.
- *Pests:* One randomized, replicated, controlled study in the Philippines found mixed effects on pests.

- *Crop growth:* Two of three studies found that crop growth was inhibited by allelopathic residues, but these effects could be minimized by changing the timing of application. One study found effects varied between years.
- *Yield:* Three randomized, replicated, controlled studies compared crop yields in amended plots with weeded controls and found positive, negative and mixed effects. Three studies compared amended plots with unweeded controls, two found positive effects on yield and one found mixed effects (depending on crop type).
- *Profit:* One study found that amending soils increased profit compared to unweeded controls, but not compared to weeded controls.
- *Crops studied:* beans, cotton, maize, rice and wheat.
- *Assessment: likely to be ineffective or harmful (effectiveness 39%; certainty 47%; harms 30%).*

http://www.conservationevidence.com/actions/728

7.4 Perennial farming

Based on the collated evidence, what is the current assessment of the effectiveness of interventions on perennial farming systems for enhancing natural pest regulation?	
Likely to be beneficial	• Exclude ants that protect pests
Unknown effectiveness (limited evidence)	• Allow natural regeneration of ground cover beneath perennial crops • Isolate colonies of beneficial ants

Likely to be beneficial

Exclude ants that protect pests

- *Parasitism:* One of two replicated, controlled studies (one also randomized) from Japan and the USA found greater parasitism of pests by natural enemies when ants were excluded from trees. The other study found greater parasitism at one site but no effect at another.
- *Natural enemies:* Five studies (including four randomized, replicated, controlled trials) from Japan, Switzerland and the USA found effects varied between natural enemy species and groups, sampling dates, sites, crop varieties and ground cover types beneath trees.
- *Pests:* Three of seven studies (including four randomized, replicated, controlled trials) found fewer pests and another found fewer pests at times of peak abundance only. One study found mixed effects depending on date and other actions taken simultaneously (predator attractant and ground cover treatments). One study found no effect.

- *Damage and tree growth:* One study found no effect on damage to tree foliage but one study found greater tree growth.
- *Ants:* Six studies found that glue or pesticide barriers reduced ant numbers in tree or vine canopies. One study found that citrus oil barriers had no effect.
- *Crops studied:* cherimoyas, cherry, grape, grapefruit, orange, pecan and satsuma mandarin.
- *Assessment: likely to be beneficial (effectiveness 40%; certainty 50%; harms 12%).*

http://www.conservationevidence.com/actions/886

Unknown effectiveness (limited evidence)

Allow natural regeneration of ground cover beneath perennial crops

- *Natural enemies on crop trees and vines:* Five studies (including one replicated, randomized, controlled test) from Australia, China, Italy and Portugal compared natural and bare ground covers by measuring numbers of natural enemies in fruit tree or vine canopies. Three found effects varied between groups of natural enemies, two found no difference. Two studies from Australia and France compared natural to sown ground cover and found no effect on enemies in crop canopies.
- *Natural enemies on the ground:* Five studies (including three replicated, randomized, controlled trials) from Australia, Canada, China, France, and Spain compared natural and bare ground covers by measuring natural enemies on the ground. Two studies found more natural enemies in natural ground cover, but in one the effects were only short-term for most natural enemy groups. Three studies found mixed effects, with higher numbers of some natural enemy groups but not others. Two studies compared natural and sown ground covers, one study found more natural enemies and one found no effect.
- *Pests and crop damage:* Four studies (three controlled, one also replicated and randomized) from Italy, Australia and China measured pests and crop damage in regenerated and bare ground

covers. Two studies found fewer pests, whilst two studies found effects on pests and crop damage varied for different pest or disease groups. One study found more pests in natural than in sown ground covers.

- *Crops studied:* apple, grape, lemon, olive and pear.
- *Assessment: unknown effectiveness (effectiveness 35%; certainty 29%; harms 20%).*

http://www.conservationevidence.com/actions/720

Isolate colonies of beneficial ants

- *Natural enemies:* One replicated, controlled study from Australia found predatory ants occupied more cashew trees when colonies were kept isolated.
- *Pest damage and yield:* The same study found lower pest damage to cashews and higher yields.
- *The crop studied* was cashew.
- *Assessment: unknown effectiveness (effectiveness 60%; certainty 19%; harms 0%).*

http://www.conservationevidence.com/actions/773

7.5 Livestock farming and pasture

Based on the collated evidence, what is the current assessment of the effectiveness of interventions on livestock and pasture farming systems for enhancing natural pest regulation?	
Likely to be beneficial	• Grow plants that compete with damaging weeds
Unknown effectiveness (limited evidence)	• Delay mowing or first grazing date on pasture or grassland
Likely to be ineffective or harmful	• Use grazing instead of cutting for pasture or grassland management • Use mixed pasture

Likely to be beneficial

Grow plants that compete with damaging weeds

- *Weed weight and cover*: Nine studies from Australia, Slovakia, the UK and the USA tested the effects of planting species to compete with weeds. All (including four replicated, randomized, controlled trials) found reduced weed plant weight or ground cover, although two found this only in some years or conditions.
- *Weed reproduction and survival:* Five studies (including three replicated, randomized, controlled trials) also found that competition reduced weed reproduction, survival or both. One of these found an effect only in one year only.

- *Crops studied*: clovers, fescues, ryegrass, other grasses and turnip.
- *Assessment: likely to be beneficial (effectiveness 70%; certainty 60%; harms 5%).*

http://www.conservationevidence.com/actions/722

Unknown effectiveness (limited evidence)

Delay mowing or first grazing date on pasture or grassland

- *Natural enemy abundance:* One replicated, randomized, controlled study found fewer predatory spiders with delayed cutting. Three studies from the UK (two of them replicated, randomized and controlled) found no change in insect predator numbers and one replicated study from Sweden found mixed effects between different predator groups.
- *Natural enemy diversity:* One replicated study from Sweden found a decrease in ant diversity with delayed cutting and one replicated, randomized, controlled study from the UK found no effect on spider and beetle diversity.
- *Pests:* One of two replicated, randomized, controlled studies from the UK and USA found more pest insects in late-cut plots and one found no effect.
- *Insects in general:* Four replicated, randomized, controlled studies measured the abundance of insect groups without classifying them as pests or natural enemies. One UK study found lower numbers in late-cut plots, while two found effects varied between groups. Two studies from the UK and USA found no effect on insect numbers.
- *Crops studied:* barley, bird's-foot trefoil, clovers, fescues, rapeseed, ryegrass, other grasses and wheat.
- *Assessment: unknown effectiveness (effectiveness 5%; certainty 20%; harms 15%).*

http://www.conservationevidence.com/actions/727

Likely to be ineffective or harmful

Use grazing instead of cutting for pasture or grassland management

- *Natural enemies:* Two studies (one before-and-after and one replicated trial) from Australia and the UK found grazing instead of cutting had mixed effects on natural enemies, with some species and groups affected on some dates but not others. One replicated study from New Zealand found no effect.
- *Pests and diseases:* One of five studies (including three replicated trials) from Australia, New Zealand, the UK and the USA found more pests, and two studies found effects varied between pest groups and sampling dates. Two studies found no effect on pests. One study found no effect on disease when grazing was used in addition to cutting.
- *Pasture damage and plant survival:* One randomized study found more ryegrass shoots were attacked by pests. One study found lower survival of alfalfa plants but another found no effect.
- *Yield:* One of four randomized, replicated studies (one also controlled) found lower yields and two found no effect. One study found lower ryegrass and higher clover yields, but no difference between clover varieties. Another randomized study found more ryegrass shoots.
- *Crops studied:* alfalfa, cock's-foot, perennial ryegrass, other grasses and white clover.
- *Assessment: likely to be ineffective or harmful (effectiveness 10%; certainty 45%; harms 40%).*

http://www.conservationevidence.com/actions/885

Use mixed pasture

- *Weeds:* Two of two studies (randomized and replicated and one also controlled) from the USA found weeds were negatively affected by mixed compared to monoculture pasture.

- *Pests:* Five studies from North America measured pests including four randomized, replicated, controlled tests. One study found fewer pests and two studies found negative or mixed effects depending on different pests groups or pasture mixes. One study found no effect ad another found more pests, although the effect was potentially inseparable from grazing treatments.
- *Crop mortality:* One randomized, replicated study from the USA found no effect on forage crop mortality caused by nematodes.
- *Yield:* Two of five studies (including two randomized, replicated, controlled tests) from North America found increased forage crop yields and two studies found mixed effects depending on the crop type and year. One study found no effect.
- *Crops studied:* alfalfa, bird's-foot trefoil, chicory, cicer milkvetch, clovers, fescues, oats, plantain, ryegrass, other grasses, other legumes, rapeseed and turnip.
- *Assessment: likely to be ineffective or harmful (effectiveness 35%; certainty 45%; harms 20%).*

http://www.conservationevidence.com/actions/721

8. ENHANCING SOIL FERTILITY

Georgina Key, Mike Whitfield, Lynn V. Dicks, William J. Sutherland & Richard D. Bardgett

Expert assessors

Martin Collison, Collison and Associates Limited, UK
Julia Cooper, Newcastle University, UK
Thanasis Dedousis, PepsiCo Europe
Richard Heathcote, Heineken, S&N UK Ltd
Shamal Mohammed, Agriculture and Horticulture Development Board, Cranfield University, UK
Andrew Molyneux, Huntapac Produce Ltd, UK
Wim van der Putten, Netherlands Institute of Ecology
Brendan Roth, Department for Environment, Food & Rural Affairs, UK
Franciska de Vries, University of Manchester, UK

Scope of assessment: actions to enhance soil fertility for agricultural systems across the world.

Assessed: 2014.

Effectiveness measure is the median % score.

Certainty measure is the median % certainty of evidence, determined by the quantity and quality of the evidence in the synopsis.

Harm measure is the median % score for negative side-effects for the farmer such as reduced yield, crop quality or profits, or increased costs.

 http://dx.doi.org/10.11647/OBP.0109.08

This book is meant as a guide to the evidence available for different conservation interventions and as a starting point in assessing their effectiveness. The assessments are based on the available evidence for the target group of species for each intervention. The assessment may therefore refer to different species or habitat to the one(s) you are considering. Before making any decisions about implementing interventions it is vital that you read the more detailed accounts of the evidence in order to assess their relevance for your study species or system.

Full details of the evidence are available at
www.conservationevidence.com

There may also be significant negative side-effects on the target groups or other species or communities that have not been identified in this assessment.

A lack of evidence means that we have been unable to assess whether or not an intervention is effective or has any harmful impacts.

8.1 Reducing agricultural pollution

Based on the collated evidence, what is the current assessment of the effectiveness of interventions to reduce agricultural pollution for enhancing soil fertility?	
Unknown effectiveness (limited evidence)	• Change the timing of manure application
Likely to be ineffective or harmful	• Reduce fertilizer, pesticide or herbicide use generally

Unknown effectiveness (limited evidence)

Change the timing of manure application

- One controlled, randomized, replicated, site comparison study from the UK found less nitrate was lost from the soil when manure application was delayed from autumn until December or January.
- *Soil types covered*: sandy loam.
- *Assessment: unknown effectiveness (effectiveness 50%; certainty 33%; harms 24%).*

http://www.conservationevidence.com/actions/893

Likely to be ineffective or harmful

Reduce fertilizer, pesticide or herbicide use generally

- *Biodiversity*: Two site comparison studies from Italy and Pakistan (one also replicated) found a higher diversity of soil invertebrates and microorganisms in low chemical-input systems.
- *Nutrient loss*: One study from Canada found lower nutrient levels and yields in low-input systems.
- *Soil types covered*: course sandy, loam, sandy loam, and silt.
- *Assessment: likely to be ineffective or harmful (effectiveness 26%; certainty 40%; harms 48%).*

http://www.conservationevidence.com/actions/904

8.2 All farming systems

Based on the collated evidence, what is the current assessment of the effectiveness of interventions on all farming systems for enhancing soil fertility?	
Likely to be beneficial	• Control traffic and traffic timing
Trade-off between benefit and harms	• Change tillage practices • Convert to organic farming • Plant new hedges
Unknown effectiveness (limited evidence)	• Change the timing of ploughing

Likely to be beneficial

Control traffic and traffic timing

- *Biodiversity*: One randomised, replicated study from Poland found higher numbers and bacterial activity under controlled traffic. One replicated site comparison study from Denmark found higher microbial biomass when farm traffic was not controlled.
- *Erosion*: Five trials from Europe and Australia (including three replicated trials, one controlled before-and-after trial, and one review) found a higher number of pores in the soil, less compaction, reduced runoff and increased water filtration into soil under controlled traffic. One controlled, replicated trial in India found increased soil crack width when traffic was not controlled.

- *Yield*: One replicated trial from Australia found increased yield under controlled traffic.
- *Soil types covered*: clay, loamy silt, sandy loam, silty, silty clay, silt-loam.
- *Assessment: likely to be beneficial (effectiveness 55%; certainty 62%; harms 18%).*

http://www.conservationevidence.com/actions/899

Trade-off between benefit and harms

Change tillage practices

- *Biodiversity loss*: Nine studies from Canada, Europe, Mexico, or the USA measured effects of reduced tillage on soil animals or microbes. Of these, six (including three replicated trials (two also randomized and one also controlled) found more microbes, more species of earthworm, or higher microbe activity under reduced tillage. One replicated trial found increased numbers of soil animals and earthworms under reduced tillage. Two (including one controlled, replicated trial), found no effect of reduced tillage on earthworm activity or microbe activity.
- *Compaction*: Five studies from Australia, Canada, and Europe measured the effect of controlled traffic and reduced tillage on compacted soils. Of these, two (including one before-and-after trial and one replicated trial) found reduced compaction and subsequent effects (reduced water runoff, for example) under controlled traffic, and one also found that crop yields increased under no-tillage. Three replicated trials, including one site comparison study, found higher compaction under reduced tillage.
- *Drought*: Three replicated trials from Europe and India (one randomized) found the size of soil cracks decreased, and ability of soil to absorb water and soil water content increased with conventional tillage and sub-soiling.
- *Erosion*: Ten replicated trials from Brazil, Europe, India, Nigeria and the USA, and one review showed mixed results of tillage on soil

erosion. Seven trials (one also controlled and randomized) showed reduced soil loss and runoff under reduced tillage compared to conventional ploughing. One trial showed no differences between tillage systems, but demonstrated that across-slope cultivation reduced soil loss compared to up-and-downslope cultivation. Two trials, showed that no-tillage increased soil loss in the absence of crop cover.

- *Soil organic carbon*: Twelve studies from Australia, Canada, China, Europe, Japan and the USA compared the effect of no-tillage and conventionally tilled systems on soil organic carbon. All (including two randomized, five replicated, two randomized, replicated, and one controlled, randomized, replicated) found higher soil organic carbon in soils under a no-tillage or reduced tillage system compared to conventionally tilled soil. One review showed that no-tillage with cover cropping plus manure application increased soil organic carbon. One randomized, replicated trial from Spain found greater soil organic carbon in conventionally tilled soil.
- *Soil organic matter*: Twelve studies from Canada, China, Europe, Morocco, and the USA measured effects of reduced tillage on soil organic matter content and nutrient retention. Of these, six studies (including three replicated, two site comparisons (one also replicated) and one controlled) found maintained or increased soil organic matter and improved soil structure under reduced tillage. Four trials (including two replicated and two site comparison studies) found higher nutrient retention under reduced tillage. One controlled, replicated trial from the USA found less carbon and nitrate in no-till compared to conventionally tilled soil, but conventionally tilled soil lost more carbon and nitrate.
- *Soil types covered*: anthrosol, calcareous silt-loam, chalky, clay, clay-loam, fine sandy loam, loam, loamy clay, loam/sandy loam, loam silt-loam, loamy silt, non-chalky clay, sandy, sandy clay-loam, sandy loam, sandy silt-loam, silt-loam, silty, silty clay, silty clay-loam, silty loam.
- *Assessment: trade-offs between benefit and harms (effectiveness 61%; certainty 72%; harms 46%).*

http://www.conservationevidence.com/actions/906

Convert to organic farming

- *Biodiversity*: Four studies in Asia, Europe, and the USA (including two site comparison studies and three replicated trials) found higher numbers, diversity, functional diversity (see background) or activity of soil organisms under organic management.
- *Soil organic carbon*: Two replicated trials in Italy and the USA showed that organically managed orchards had higher soil carbon levels compared to conventionally managed orchards. One randomised, replicated trial in the USA found soil carbon was lower under organic management compared to alley cropping.
- *Soil organic matter*: One replicated trial in Canada found that soil nutrients were lower in organically managed soils.
- *Yield*: One replicated trial in Canada found lower yields in organically managed soils. Two replicated trials in the USA (one also randomised) found that fruit was of a higher quality and more resistant to disease, though smaller or that organic management had mixed effects on yield.
- *Soil types covered*: clay, clay-loam, fine sandy loam, loam, sandy loam, sandy clay-loam, silt, silty clay, silt-loam.
- *Assessment: trade-offs between benefit and harms (effectiveness 55%; certainty 52%; harms 64%).*

http://www.conservationevidence.com/actions/895

Plant new hedges

- Five studies in Slovakia, Kenya and Thailand measured the effects of planting grass or shrub hedgerows on soil animals and soil fertility. All five found hedgerows to maintain or improve soil fertility and soil animal activity. Of these, three replicated studies found reduced soil erosion and higher soil organic matter levels. Another replicated trial found a higher diversity of soil animals near to the hedgerows. One of the replicated studies and one review found that adding woody species to the hedgerows improved many factors contributing to soil fertility.
- *Soil types covered*: alluvial, clay, sandy loam.

- *Assessment: trade-offs between benefit and harms (effectiveness 49%; certainty 45%; harms 20%).*

http://www.conservationevidence.com/actions/744

Unknown effectiveness (limited evidence)

Change the timing of ploughing

- *Nutrient loss*: Two replicated site comparison studies from Denmark and Norway (one also randomised) found reduced erosion soil loss and nitrate leaching when ploughing was delayed until spring.
- *Soil types covered*: Sandy, sandy loam, silty clay loam.
- *Assessment: unknown effectiveness (effectiveness 46%; certainty 38%; harms 33%).*

http://www.conservationevidence.com/actions/712

8.3 Arable farming

Based on the collated evidence, what is the current assessment of the effectiveness of interventions on arable farming systems for enhancing soil fertility?	
Beneficial	• Amend the soil using a mix of organic and inorganic amendments • Grow cover crops when the field is empty • Use crop rotation
Likely to be beneficial	• Amend the soil with formulated chemical compounds • Grow cover crops beneath the main crop (living mulches) or between crop rows
Trade-off between benefit and harms	• Add mulch to crops • Amend the soil with fresh plant material or crop remains • Amend the soil with manures and agricultural composts • Amend the soil with municipal wastes or their composts • Incorporate leys into crop rotation • Retain crop residues
Unknown effectiveness (limited evidence)	• Amend the soil with bacteria or fungi • Amend the soil with composts not otherwise specified • Amend the soil with crops grown as green manures • Amend the soil with non-chemical minerals and mineral wastes • Amend the soil with organic processing wastes or their composts • Encourage foraging waterfowl • Use alley cropping

Beneficial

Amend the soil using a mix of organic and inorganic amendments

- *Biodiversity*: Five controlled trials from China and India (four also randomized and replicated), and one study from Japan found higher microbial biomass and activity in soils with a mix of manure and inorganic fertilizers. Manure alone also increased microbial biomass. One trial found increased microbial diversity.
- *Erosion*: One controlled, replicated trial from India found that mixed amendments were more effective at reducing the size of cracks in dry soil than inorganic fertilizers alone or no fertilizer.
- *Soil organic carbon loss*: Four controlled, randomized, replicated trials and one controlled trial all from China and India found more organic carbon in soils with mixed fertilizers. Manure alone also increased organic carbon. One trial also found more carbon in soil amended with inorganic fertilizers and lime.
- *Soil organic matter loss*: Three randomized, replicated trials from China and India (two also controlled), found more nutrients in soils with manure and inorganic fertilizers. One controlled, randomized, replicated trial from China found inconsistent effects of using mixed manure and inorganic fertilizers.
- *Yield*: Two randomized, replicated trials from China (one also controlled) found increased maize or rice and wheat yields in soils with mixed manure and inorganic fertilizer amendments. One study found lower yields of rice and wheat under mixed fertilizers.
- *Soil types covered*: clay, clay-loam, sandy loam, silt clay-loam, silty loam.
- *Assessment: beneficial (effectiveness 69%; certainty 64%; harms 15%).*

http://www.conservationevidence.com/actions/902

Grow cover crops when the field is empty

- *Biodiversity*: One controlled, randomized, replicated experiment in Martinique found that growing cover crops resulted in more diverse nematode communities. One replicated trial from the USA found greater microbial biomass under ryegrass compared to a ryegrass/vetch cover crop mix.

- *Soil structure:* Three randomized, replicated studies from Denmark, Turkey and the UK found that growing cover crops improved soil structure and nutrient retention. One trial found higher soil porosity, interconnectivity and one lower resistance in soil under cover crops, and one found reduced nitrate leaching.
- *Soil organic carbon:* One replicated study from Denmark and one review based mainly in Japan found increased soil carbon levels under cover crops. One study also found soil carbon levels increased further when legumes were included in cover crops.
- *Soil organic matter:* Two controlled, randomized, replicated studies from Australia and the USA found increased carbon and nitrogen levels under cover crops, with one showing that they increased regardless of whether those crops were legumes or not. Two studies from Europe (including one controlled, replicated trial) found no marked effect on soil organic matter levels.
- *Yield:* One replicated trial from the USA found higher tomato yield from soils which had been under a ryegrass cover crop.
- *Soil types covered*: clay, loam, sandy clay, sandy loam, silty clay, silty loam.
- *Assessment: beneficial (effectiveness 75%; certainty 67%; harms 16%).*

http://www.conservationevidence.com/actions/898

Use crop rotation

- *Biodiversity:* Three randomized, replicated trials from Canada and Zambia measured the effect of including legumes in crop rotations and found the number of microbes and diversity of different soil animals increased.
- *Erosion:* One randomized, replicated trial from Canada found that including forage crops in crop rotations reduced rainwater runoff and soil loss, and one replicated trial from Syria showed that including legumes in rotation increased water infiltration (movement of water into the soil).

- *Soil organic carbon*: Three studies from Australia, Canada, and Denmark (including one controlled replicated trial and one replicated site comparison study), found increased soil organic carbon under crop rotation, particularly when some legumes were included.
- *Soil organic matter:* Two of four replicated trials from Canada and Syria (one also controlled and randomized) found increased soil organic matter, particularly when legumes were included in the rotation. One study found lower soil organic matter levels when longer crop rotations were used. One randomized, replicated study found no effect on soil particle size.
- *Soil types covered*: clay, clay-loam, fine clay, loam, loam/silt loam, sandy clay, sandy loam, silty loam.
- *Assessment: beneficial (effectiveness 66%; certainty 75%; harms 8%).*

http://www.conservationevidence.com/actions/857

Likely to be beneficial

Amend the soil with formulated chemical compounds

- *Nutrient loss*: Three of five replicated trials from New Zealand and the UK measured the effect of applying nitrification inhibitors to the soil and three found reduced nitrate losses and nitrous oxide emissions, although one of these found that the method of application influenced its effect. One trial found no effect on nitrate loss. One trial found reduced nutrient and soil loss when aluminium sulphate was applied to the soil.
- *Soil organic matter*: Four of five studies (including two controlled, randomised and replicated and one randomised and replicated) in Australia, China, India, Syria and the UK testing the effects of adding chemical compounds to the soil showed an increase in soil organic matter or carbon when nitrogen or phosphorus fertilizer was applied. One site comparison study showed that a slow-release fertilizer resulted in higher nutrient retention. One study found higher carbon levels when NPK fertilizers were applied with straw, than when applied alone, and one replicated study from France found higher soil carbon when manure rather than chemical compounds were applied.

- *Yield*: One replicated experiment from India showed that maize and wheat yield increased with increased fertilizer application.
- *Soil types covered*: clay, fine loamy, gravelly sandy loam, loam, sandy loam, silty, silty clay, silt-loam.
- *Assessment: likely to be beneficial (effectiveness 64%; certainty 46%; harms 19%).*

http://www.conservationevidence.com/actions/909

Grow cover crops beneath the main crop (living mulches) or between crop rows

- *Biodiversity:* One randomized, replicated study from Spain found that cover crops increased bacterial numbers and activity.
- *Erosion:* Two studies from France and the USA showed reduced erosion under cover crops. One controlled study showed that soil stability was highest under a grass cover, and one randomized replicated study found that cover crops reduced soil loss.
- *Soil organic matter:* Two controlled trials from India and South Africa (one also randomized and replicated) found that soil organic matter increased under cover crops, and one trial from Germany found no effect on soil organic matter levels.
- *Soil types covered*: gravelly sandy loam, sandy loam, sandy, silty loam.
- *Assessment: likely to be beneficial (effectiveness 65%; certainty 54%; harms 19%).*

http://www.conservationevidence.com/actions/897

Trade-off between benefit and harms

Add mulch to crops

- *Biodiversity*: Three replicated trials from Canada, Poland and Spain (including one also controlled, one also randomised and one also controlled and randomised) showed that adding mulch to crops (whether shredded paper, municipal compost or straw) increased soil animal and fungal numbers, diversity and activity. Of these, one trial also showed that mulch improved soil structure and increased soil organic matter.

- *Nutrient loss*: One replicated study from Nigeria found higher nutrient levels in continually cropped soil.
- *Erosion*: Five studies from India, France, Nigeria and the UK (including one controlled, randomised, replicated trial, one randomised, replicated trial, two replicated (one also controlled), and one controlled trial) found that mulches increased soil stability, and reduced soil erosion and runoff. One trial found that some mulches are more effective than others.
- *Drought*: Two replicated trials from India found that adding mulch to crops increased soil moisture.
- *Yield*: Two replicated trials from India found that yields increased when either a live mulch or vegetation barrier combined with mulch was used.
- *Soil types covered*: clay, fine loam, gravelly sandy loam, sandy, sandy clay, sandy loam, sandy silt-loam, silty, silty loam.
- *Assessment: trade-offs between benefit and harms (effectiveness 60%; certainty 64%; harms 23%).*

http://www.conservationevidence.com/actions/887

Amend the soil with fresh plant material or crop remains

- *Biodiversity*: One randomized, replicated experiment from Belgium found increased microbial biomass when crop remains and straw were added.
- *Compaction*: One before-and-after trial from the UK found that incorporating straw residues by discing (reduced tillage) did not improve anaerobic soils (low oxygen levels) in compacted soils.
- *Erosion*: Two randomized, replicated studies from Canada and India measured the effect of incorporating straw on erosion. One found straw addition reduced soil loss, and one found mixed effects depending on soil type.
- *Nutrient loss*: Two replicated studies from Belgium and the UK (one also controlled and one also randomized) reported higher soil nitrogen levels when compost or straw was applied, but mixed results when processed wastes were added.

- *Soil organic carbon*: Three randomized, replicated studies (two also controlled) from China and India, and one controlled before-and-after site comparison study from Denmark found higher carbon levels when plant material was added. One found higher carbon levels when straw was applied along with NPK fertilizers. One also found larger soil aggregates.
- *Soil types covered*: clay, clay-loam, loam/sandy loam, loamy sand, sandy, sandy clay-loam, sandy loam, silt-loam, silty, silty clay.
- *Assessment: trade-offs between benefit and harms (effectiveness 53%; certainty 53%; harms 34%).*

http://www.conservationevidence.com/actions/910

Amend the soil with manures and agricultural composts

- *Biodiversity loss*: Three controlled, replicated studies from the UK and USA found higher microbial biomass when manure or compost was applied, and higher microbial respiration when poultry manure was applied.
- *Erosion*: One controlled, randomized, replicated study from India found lower soil loss and water runoff with manure application in combination with other treatments.
- *Nutrient management*: Two randomized, replicated studies from Canada and the UK (one also controlled) found lower nitrate loss or larger soil aggregates (which hold more nutrients) when manure was applied, compared to broiler (poultry) litter, slurry or synthetic fertilizers. One study found that treatment in winter was more effective than in autumn and that farmyard manure was more effective than broiler (poultry) litter or slurry in reducing nutrient loss. One controlled, replicated study from Spain found higher nitrate leaching.
- *Soil organic carbon*: Three studies (including two controlled, replicated studies and a review) from India, Japan and the UK found higher carbon levels when manures were applied.
- *Soil organic matter*: One controlled, randomized, replicated study from Turkey found higher organic matter, larger soil aggregations

and a positive effect on soil physical properties when manure and compost were applied. One study from Germany found no effect of manure on organic matter levels.

- *Yield*: Four controlled, replicated studies (including four also randomized) from India, Spain and Turkey found higher crop yields when manures or compost were applied. One study found higher yields when manure were applied in combination with cover crops.
- *Soil types covered*: clay-loam, loam, loamy, sandy loam, sandy clay-loam, silty loam, sandy silt-loam.
- *Assessment: trade-offs between benefit and harms (effectiveness 70%; certainty 59%; harms 26%).*

http://www.conservationevidence.com/actions/911

Amend the soil with municipal wastes or their composts

- *Erosion*: Two controlled, replicated trials in Spain and the UK measured the effect of adding wastes to the soil. One trial found that adding municipal compost to semi-arid soils greatly reduced soil loss and water runoff. One found mixed results of adding composts and wastes.
- *Soil types covered*: coarse loamy, sandy loam.
- *Assessment: trade-offs between benefit and harms (effectiveness 45%; certainty 44%; harms 54%).*

http://www.conservationevidence.com/actions/890

Incorporate leys into crop rotation

- *Nutrient loss*: One replicated study from Denmark showed that reducing the extent of grass pasture in leys reduced the undesirable uptake of nitrogen by grasses, therefore requiring lower rates of fertilizer for subsequent crops.
- *Soil types covered*: sandy loam.
- *Assessment: trade-offs between benefit and harms (effectiveness 46%; certainty 45%; harms 36%).*

http://www.conservationevidence.com/actions/900

Retain crop residues

- *Biodiversity:* One replicated study from Mexico found higher microbial biomass when crop residues were retained.
- *Erosion*: One review found reduced water runoff, increased water storage and reduced soil erosion. One replicated site comparison from Canada found mixed effects on soil physical properties, including penetration resistance and the size of soil aggregates. One replicated study from the USA found that tillage can have mixed results on soil erosion when crop remains are removed.
- *Soil organic matter:* One randomized, replicated trial from Australia found higher soil organic carbon and nitrogen when residues were retained, but only when fertilizer was also applied.
- *Yield:* One randomized, replicated trial from Australia found higher yields when residues were retained in combination with fertilizer application and no-tillage.
- *Soil types covered:* clay, loam, sandy loam, silt-loam.
- *Assessment: trade-offs between benefit and harms (effectiveness 63%; certainty 54%; harms 29%).*

http://www.conservationevidence.com/actions/907

Unknown effectiveness (limited evidence)

Amend the soil with bacteria or fungi

- *Biodiversity*: One randomised, replicated trial from India showed that adding soil bacteria and arbuscular mycorrhizal fungi resulted in higher microbial diversity.
- *Soil organic matter*: One controlled, randomised, replicated trial from Turkey found increased soil organic matter content in soil under mycorrhizal-inoculated compost applications.
- *Yield*: Two randomised, replicated trials (including one also controlled) from India and Turkey found higher crop yields.
- *Soil types covered*: clay-loam, sandy loam.
- *Assessment: unknown effectiveness (effectiveness 40%; certainty 31%; harms 17%).*

http://www.conservationevidence.com/actions/888

Amend the soil with composts not otherwise specified

- *Soil organic matter*: One controlled, randomised, replicated trial in Italy found that applying a high rate of compost increased soil organic matter levels, microbial biomass and fruit yield.
- *Soil types covered*: silty clay.
- *Assessment: unknown effectiveness (effectiveness 54%; certainty 29%; harms 19%).*

http://www.conservationevidence.com/actions/889

Amend the soil with crops grown as green manures

- *Soil organic matter*: Two controlled, randomized, replicated studies from India and Pakistan found higher soil organic carbon, and one found increased grain yields when green manures were grown.
- *Soil types covered*: clay-loam.
- *Assessment: unknown effectiveness (effectiveness 53%; certainty 36%; harms 16%).*

http://www.conservationevidence.com/actions/908

Amend the soil with non-chemical minerals and mineral wastes

- *Nutrient loss*: Two replicated studies from Australia and New Zealand measured the effects of adding minerals and mineral wastes to the soil. Both found reduced nutrient loss and one study found reduced erosion.
- *Soil types covered*: sandy clay, silt-loam.
- *Assessment: unknown effectiveness (effectiveness 35%; certainty 37%; harms 23%).*

http://www.conservationevidence.com/actions/892

Amend the soil with organic processing wastes or their composts

- *Nutrient loss*: Two controlled, replicated trials from Spain and the UK (one also randomized) measured the effect of adding composts to soil. One trial found applying high rates of cotton gin compost and poultry manure improved soil structure and reduced soil loss, but increased nutrient loss. One trial found improved nutrient retention and increased barley *Hordeum vulgare* yield when molasses were added.

- *Soil types covered*: sandy clay, sandy loam, silty clay.
- *Assessment: unknown effectiveness (effectiveness 58%; certainty 35%; harms 20%).*

http://www.conservationevidence.com/actions/891

Encourage foraging waterfowl

- *Soil organic matter:* One controlled, replicated experiment from the USA found increased straw decomposition when ducks were allowed to forage.
- *Soil types covered:* silty clay.
- *Assessment: unknown effectiveness (effectiveness 14%; certainty 34%; harms 20%).*

http://www.conservationevidence.com/actions/711

Use alley cropping

- *Biodiversity*: A controlled, randomized, replicated study from Canada found that intercropping with trees resulted in a higher diversity of arbuscular mycorrhizal fungi.
- *Soil types covered*: sandy loam.
- *Assessment: unknown effectiveness (effectiveness 36%; certainty 23%; harms 19%).*

http://www.conservationevidence.com/actions/903

8.4 Livestock and pasture farming

Based on the collated evidence, what is the current assessment of the effectiveness of interventions on livestock and pasture farming systems for enhancing soil fertility?	
Likely to be beneficial	• Reduce grazing intensity
Trade-off between benefit and harms	• Restore or create low input grasslands

Likely to be beneficial

Reduce grazing intensity

- *Compaction*: One replicated study from Australia found compacted soils recovered when sheep were excluded for 2.5 years.
- *Erosion*: Two replicated studies from New Zealand, and Syria (one also controlled) measured the effect of grazing animals on soil nutrient and sediment loss. Of these, one trial found increased soil carbon and nitrogen when grazing animals were excluded. One trial found higher soil phosphate levels, and less sediment erosion when grazing time in forage crops was reduced.
- *Soil types covered*: clay, clay-loam, loamy, silt-loam.
- *Assessment: likely to be beneficial (effectiveness 51%; certainty 58%; harms 14%).*

http://www.conservationevidence.com/actions/901

Trade-off between benefit and harms

Restore or create low input grasslands

- *Biodiversity*: One randomized, replicated trial in the Netherlands and one controlled trial from France found that restoring grasslands increased the diversity of soil animals. One trial also found higher microbial biomass, activity and carbon under grassland.
- *Soil types covered*: sandy loam, silty.
- *Assessment: trade-offs between benefit and harms (effectiveness 53%; certainty 59%; harms 32%).*

http://www.conservationevidence.com/actions/905

www.ingramcontent.com/pod-product-compliance
Lightning Source LLC
LaVergne TN
LVHW082108200926
848484LV00008B/1929